PROBLEM SOILS
CONSTRAINTS AND MANAGEMENT

This is a unique book that deals with the problem soils, their constraints and management in the Indian context. The book starts with the introduction on problem soils and the classification of these soils are included there under. In India, there is wide spread occurrence of soils with different types of constraints for crop production. Such soils are popularly called as "Problem Soils". Cultivation in these soils is not so easy as several problems have to be tackled during cultivation. It may be either soil droughtiness or acidity or salinity etc. An attempt has been made in this book to cover most of the problematic soils in India. The classification of problem soils has been done based on the limitations they possess and the most dominant limitation is taken into consideration for grouping it under a particular class. Here five broader classes have been identified viz., soils with climatic problems; soils with physical problems; soils with chemical problems; soils with biological problems and soils with problems due to anthropogenic reasons.

Dr. K C Manorama Thampatti is the Head, Department of Soil Science & Agricultural Chemistry, College of Agriculture, Vellayani, Kerala Agricultural University, Thiruvananthapuram, Kerala. She is a faculty of the Kerala Agricultural University since 1987. Her fields of research specialization are wet land chemistry, soil and water pollution assessment and mitigation, bioremediation of soil contaminants, and solid waste management. She has to her credit more than 50 research papers in peer reviewed international and national journals and three chapters in books.

PROBLEM SOILS

CONSTRAINTS AND MANAGEMENT

K.C. Manorama Thampatti
Department of Soil Science & Agricultural Chemistry
College of Agriculture
Kerala Agricultural University
Vellayani, Thiruvananthapuram
KERALA-695522

CRC Press is an imprint of the
Taylor & Francis Group, an **informa** business

NARENDRA PUBLISHING HOUSE
DELHI (INDIA)

First published 2023
by CRC Press
4 Park Square, Milton Park, Abingdon, Oxon, OX14 4RN

and by CRC Press
6000 Broken Sound Parkway NW, Suite 300, Boca Raton, FL 33487-2742

CRC Press is an imprint of Informa UK Limited

British Library Cataloguing-in-Publication Data
A catalogue record for this book is available from the British Library

ISBN: 9781032388977 (hbk)
ISBN: 9781032388984 (pbk)
ISBN: 9781003347354 (ebk)

DOI: 10.1201/9781003347354

Typeset in Arial, Britannica, Calibri, Cambria, Symbol and Times New Roman
by Amrit Graphics, Delhi 110032

CONTENTS

PREFACE

I have prepared this text book based on the knowledge, I have acquired through my teaching and research. Of course, whatever I have studied from my teachers have thrown light on the subject. Whenever I go through various literature on problem soils, I always feel that a comprehensive text book on problem soils is lacking, especially that will be much useful to graduate and postgraduate students. I always thought of "why nobody is publishing a unique text on Problem Soils and their Management". I am not forgetting the few books that are addressing these matters. So, I finally decided to document my knowledge on problem soils in a systematic way and that gave birth to this publication.

Problem soils are the ones that have serious constraints for crop production and that need special management techniques. Generally, these soils posses certain characteristics that make them uneconomical for the cultivation of crops without adopting proper reclamation measures and may be natural or anthropogenic. For successful utilization of these soils, their limitations have to be identified and suitable corrective measures have to be adopted.

Classification of problem soils based on their limitations will help to reorient their management strategies. A broader classification which could accommodate the existing soil problems and that will arise in future is a better option. An attempt in that line is presented in this book. Major five groups, have been suggested which could accommodate most of the problematic soils in one group or another. The broader groups suggested are soils with climatic problems, soils with physical problems, soils with chemical problems, soils with biological problems and that with problems due to anthropogenic reasons or degraded soils.

If a soil can be included under two groups, it will be included under the group based on the most dominant problem. Thus, this book deals with 23 types of problem soils and their characteristics and corrective measures giving due importance to all the problems. A little more emphasis is being given to submerged soils and salt affected soils.The text is prepared exclusively based on the Indian context.

Hope that the book will meet the growing needs of the undergraduate students of agriculture as well as the post graduate students and teachers of Soil Science and Agricultural Chemistry. This is my first attempt and I welcome any suggestion on the matter presented.

The author expresses her sincere thanks to Dr. P. Sureshkumar, Dean, School of Agricultural Science, Amritha Vidyapeetham, Coimbatore, Tamil Nadu for his critical scrutiny of the manuscript.

Ms. Jaya Publishing House, Delhi deserves a special word of appreciation for the pains taken in printing this book.

K.C. Manorama Thampatti

CHAPTER 1

PROBLEM SOILS: AN INTRODUCTION

All soils in general require good management to remain productive over the long term. There are certain soils that possess characteristics that pose problems for their optimal use. Soils that have serious constraints for crop production and that need special management techniques are called problem soils. They can be defined as the soils which possess characteristics that make them uneconomical for the cultivation of crops without adopting proper reclamation measures. The problems arise either due to natural reasons or anthropogenic as in the case of degraded soils where unwise management intervention being the main culprit. The limitations may be physical, chemical or biological. Based on the limitations, appropriate management techniques are to be selected. Examples for some of the common problems are wetness, dryness, salinity, acidity, sodicity, steepness, structural limitations etc. The problems are very much specific as in the case of halophytes, where salinity is not a problem.

The problem soils have been classified mainly based on their constraints. The most discussed classes are the acid soils, salt affected soils, laterite soils, acid sulphate soils and peat soils. The Asian Network on Problem Solis has grouped problem soils into **11** classes as given below (FAOAGL, 2000).

1. Cold soils
2. Dry soils

3. Steep soils
4. Shallow soils
5. Poorly drained soils
6. Coarse textured soils (Sandy soils)
7. Heavy cracking soils (Vertisols)
8. Poorly fertile soils
9. Salt affected soils –saline, sodic, saline-sodic
10. Acid sulphate soils
11. Peat soils (Organic soils).

The above classification did not include all the problem soils. Mine soils, eroded soils, compacted soils, gravelly soils etc. are not included here. Hence a broader classification which could accommodate the existing soil problems and that will arise in future is a better option. In that way, these soils can be classified based on their most distinct problem in a broader sense. Thus, based on the nature of problems five groups are identified under this classification.

1. Soils that face climatic problems
2. Soils with physical problems
3. Soils with chemical problems
4. Soils with biological problems
5. Soils that face problems due to anthropogenic reasons

Under this classification sometimes, a soil can be grouped under two or more classes. For example, "poorly aerated soils" can be grouped under classes 2, 3 or 4. Hence the soils are included under each class based on their most distinct problem that make crop growth most difficult. The broad classes include sub classes based on their individual problems. The following table illustrates the grouping of problem soils.

Table 1. Classification of problem soils based on the nature of problems

Nature of problem				
Climatic problems	**Physical problems**	**Chemical problems**	**Biological problems**	**Anthropogenic problems**
1. Cold soils	1. Steep soils	1. Acid soils	1. Low carbon soils	1. Degraded soils
2. Dry soils	2. Shallow soils	2. Acid sulphate soils	2. Soils with low biological activity	*1.1. Eroded soils*
	3. Imperfectly drained / submerged soils	3. Laterite and associated soils		*1.2. Nutritionally poor soils*
	4. Surface crusted soils	4. Red soil		*1.3. Waterlogged soils*
	5. Soils with hardpan	5. Salt affected soils		*1.4. Contaminated soils*
	6. Highly permeable soils			*1.5. Mined degraded soils*
	7. Fluffy paddy soils			
	8. Heavy clay soils			
	9. Organic soils			

CHAPTER

SOILS WITH CLIMATIC PROBLEMS 2

Climate is one of the major factors that decides and regulates cultivation. At the same time, it is one of the prime factors that decide soil properties also. The crop choices and time of cultivation are mainly controlled by the climate. Extreme coldness and dryness are the two climatic conditions that make the plant growth miserable. Two problem soils are included under this class viz., cold soils and dry soils.

1. COLD SOILS (FROZEN SOILS)

Cold soils occupy nearly one third of the total land area of the world. They are generally found in cold climates, but much variation was observed in temperature and rainfall pattern among the cold soils found in different parts of the world. Hence wide variation is observed within temperature and moisture regimes. These soils are generally found in temperate and high altitudes areas (Himalayas, Alps, Rocky Mountains, high altitude areas of Japan and China) where the mean temperature for a day will be $<5^{o}C$ during the growing period. They are mainly skeletal soils and calcareous in nature. Natural vegetation comprises sparse forests and cultivation is very much limited. Temperate fruit trees like apple, apricot and cool season vegetables are cultivated.

There is no cultivation in areas like Artic /Antarctica. The prevailing soil conditions create a harsh environment for plant and animal life. In such situations lichens or algae are found beneath the ice caps. Only a few plants and animals have managed to survive the cryic situations.

In India, such soils are found in north-western Himalayas, covering Ladakh and Gilgit districts with an area of 15.2 M ha. This has been classified under Agro-ecoregion 1 (AER 1) - Cold Arid Ecoregion. Two sub-regions had been recognized under this as

1. Cold hyper-arid eco-subregion covering eastern parts of Ladakh Plateau and
2. Cold to cool typic-arid eco-subregion covering western parts of Ladakh Plateau.

Agroclimate is characterized by mild summer and severe winter with mean annual temperature of less than 8°C and mean annual rainfall of less than 150 mm. The precipitation is less than 15 per cent of the mean annual PET (potential evapotranspiration) and hence qualifies for aridic soil moisture and cryic soil temperature regime. The growing period available is less than 90 days.

Loamy skeletal soils, calcareous in nature occur on gentle slops to almost level valleys. They are alkaline in reaction and low to medium in organic matter content. They are classified under Great Groups of Cryorthents and Cryorthids. The area is dominantly represented by Ladakh series, classified as Typic Cryorthent. The higher northern part of the plateau remains under permanent snow cover (Gajbhiye and Mandal, 2006).

The native vegetation is mainly sparse forest trees. Vegetables, millets, wheat, fodder, pulses and barley are cultivated, but the production per unit area is low. Apple and apricot are the major fruit crops grown in the area. Among the livestock, mule dominates, while sheep, goat and yak stand next in order. This zone is known for grazing (by pashmina goats).

Constraints

The limitations are due to the harsh climate. In cryic situations weathering will be at a very low rate and soil development is very slow. The soils formed will be of poor fertility. The major limitations are the following.

1. Climatic limitation – Since the prevalence of cold situation, such soils always have cryic temperature regime (0-8 ^{O}C) resulting a very low soil temperature.
2. Narrow crop growing period – During the major part of the year the soils are snow clad and the summer season when the snow melts is the only season that permits cultivation. The cultivation season or growing period is very short. Crop selection is to be done based on the length of growing period.
3. Extreme coldness reduces the rate of weathering and the soils formed are mainly shallow, sandy and gravelly / boulder in nature
4. Majority of these soils are moderately to highly calcareous in nature.
5. Moderately to highly calcareous nature of soil results nutrient imbalance with low availability of major and micronutrients. The biological activity is also very low.

Management

In extreme cold situation, cultivation is not possible. In other areas the cultivation season is limited to summer periods. In such areas cultivation is possible by appropriate management practices as pointed out below.

1. Adjust planting time to safeguard the crops from extremes of climatic vagaries
2. Choose cold tolerant plants
3. Follow soil temperature regulating mechanisms like mulching
4. Choose protected cultivation
5. Follow scientific crop management practices

2. DRY SOILS / ARID SOILS

Dry soils are found in arid and semi-arid areas with dry climate, where the water deficit prevails throughout the year. These tracts are characterized by very low precipitation, high evaporation rates that exceeds precipitation and wide temperature swings both daily and seasonally. The aridity index (AI) will be less than one.

$$AI = \frac{\text{Precipitation}}{\text{Potential evapotranspiration}}$$

Dry climates are found throughout the globe, particularly in western North America, Australia, southern South America, central and southern Africa and much of Asia. The natural vegetation is mainly sparse sporadic thorn forest. Dry land ecosystems are crucially important, both to the environment and humans. It covers 41 per cent of earth's surface and are home to about 38 per cent of world's people (Pravalie, 2016). They also harbour a rich and unique diversity of species, and play a critical role in the global carbon cycle. Currently the drylands are expanding due to anthropogenic activities and the most recent climate forecasts predict further increase in the extent of drylands.

Weathering is slow in dry soils. Physical weathering leaves big pieces of rocks which can be further broken down by chemical weathering which is very slow due to the water deficit. Soil formation is slowed down by the low content of organic matter and scanty vegetative cover. The very little organic matter that is present in these soils will quickly lose when cultivated.

Extreme dryness will result cracking of soils and destroys soil structure. In such situations root depth is very shallow and crop growth is poor. Severe dry conditions persist during the growing period. These areas experience arid agroclimate with hot summer and cool winters. General cropping season is limited to three months. The annual rain fall is low and potential evapotranspiration is very high. Soils showed wide variation in their characteristics and are grouped under mainly soil orders of Entisols and Vertisols. These soils are less suitable for agriculture. However, rainfed monocropping is generally practiced with drought tolerant millets and fodders.

Salt accumulation is also high under dry conditions. Gypsiferous and calcareous soils are found in the driest areas on the earth, where the layers of gypsum or calcium carbonate prevent agriculture. The potential productivity of calcareous soils is higher compared to gypsiferous soils, if adequate water and nutrients are supplied. The high calcium saturation tends to keep the calcareous soils in well aggregated form and good physical condition. However, if soil contains an impermeable hardpan either calcareous or gypsiferous that should be broken before cultivation. Soils with gypsic horizon are prone to formation of sink holes when irrigated.

India has dry soils at hot arid ecoregions with desert and saline soils at Kachchh and Kathiwar Peninsula (AER 2) and red and black soils in Deccan Plateau (AER 3). Hot semi-arid ecoregions with alluvium derived soils (AER 4), medium to deep black soils (AER 5), shallow to medium black soils (AER 6), red and black soils (AER 7) and red soils (AER8) distributed at Gujarat, Northern Plains, Central Highlands and Deccan plateau also have dry soils (AER 4, 5 and 6). These soils are classified under soil orders Entisols, Aridisols, Inceptisols,Vertisols and Alfisols (Gajbhiye and Mandal, 2006).

In Vertisols (black cotton soils), the dryness makes it extremely difficult for workability due to very narrow range of workable soil moisture. Similarly subsoil sodicity, poor drainage and low oxygen availability are also common in black soils. Red loamy soils of Inceptisols and Alfisols, face very high subsoil density, limiting the rooting depth. In both soils, crop failures are common due to soil dryness.

The cropping pattern generally followed in arid ecoregions is rainfed farming. In AER 2 (Western Plain, Kachchh and part of Kathiawar Peninsula), drought resistant and short duration rainy season crops like pearl millet, fodder and pulses are grown in non-saline areas. In AER 3

(Deccan plateau) also traditional rainfed farming is practiced where the land is kept fallow during rainy season and cultivation is done during post-rainy season on residual soil moisture. The common post-rainy season crops are sorghum and safflower, and rarely pearl millet. In areas where irrigation is possible, groundnut, sunflower, sugarcane and cotton are cultivated. The crop yield is very low.

As the climate changes from hot arid to semi arid the cultivable situations are better and constraints are less. Crops like sorghum, pearl millet, maize, groundnut, pigeon pea, soyabean, pulses, cotton, rice, castor etc. are cultivated.

The information on soil biology of dry soils is scanty. However, drylands provide habitat for species uniquely adapted to variable and extreme environments. Dryland species range from micro-organisms, to ants, grasshoppers, and snakes to large carnivores such as cheetahs and leopards.

Constraints

1. Erratic and scanty rainfall: Dry climate experiencing areas usually receive scanty rainfall and that too erratic. This may lead to high water deficit in soils, making crop cultivation very difficult.
2. Salinity: Another problem is soil salinity which is very common in dry tracts. Absence of rain and high rate of evaporation causes the accumulation of salts in the surface soil often leading to physiological drought.
3. Subsoil acidity: Acute subsoil acidity is observed in soil orders Inceptisols and Alfisols
4. Acute dryness: It affects the plant adversely and such acute dryness at the time of grain formation results very low yield.
5. Nutrient imbalance: Nutrient imbalance is highly prevalent in dry soils since proper soil weathering is absent in such climates though they are inherently fertile as they contain large amounts of weatherable

minerals. These soils are deficient in N, P, Zn and Fe but are rich in bases like Ca and Mg.

Management

1. Provision of mulches, either natural or artificial to reduce evaporation loss and regulate soil temperature. Mulching provides a congenial soil environment for root growth.
2. Addition of organic manure to improve soil structure and to facilitate better soil moisture retention.
3. Adoption of scientific irrigation practices like drip or sprinkler irrigation and / or subsoil irrigation which will prevent unnecessary loss of water and improve water use efficiency.
4. Follow conservation tillage for better moisture retention and water use by crops.
5. Judicious use of soil amendments / conditioners and fertilizers to address the problems of salt accumulation, subsoil acidity and nutritional imbalance.
6. Cultivate drought tolerant crops
7. In salt affected areas, an effective drainage system has to be provided to maintain a relatively low water-table and keep salinity under control.

CHAPTER

SOILS WITH PHYSICAL PROBLEMS

3

1. STEEP SOILS

Steep soils are found in areas where percentage slope is >30. In India such soils are found mainly in Eastern Himalayas, West Bengal, northern parts of Assam, Arunachal Pradesh, Sikkim and Western Ghats in Kerala, Karnataka and Maharashtra. In many of the Agroecological regions, steep soils are found except that are dominated by alluvial soils. The natural vegetation is pine forests in northern tracts. The annual rain fall is >2000 mm. These soils rarely experience water stress, but the soils are very shallow making them very prone to erosion and landslides. In the tropics most of the steep land areas are settled by small-scale farming families where livelihoods may be endangered by land degradation and associated loss of productivity. Crops like tea, apple, pear, rubber, pineapple etc. are cultivated after taking necessary soil conservation measures.

Constraints

1. Steep slopes: Since the steepness is very high these soils are subjected to heavy run off
2. Soil erosion: Erosion hazards are very high due to steep slope and shallow depth of soil
3. Poor soil fertility: Soil are very poor in fertility due to severe leaching of nutrients

4. Shallow depth: Soil depth is very low making the root growth very poor resulting in very low crop yield
5. Decline in soil fertility and increase in soil acidity: Decline in soil fertility in steep soils is indeed a major consequence of erosion resulting from the loss of fertile surface soil. The loss of easily soluble basic cations results in the accumulation of less soluble iron and aluminium which on oxidation enhances soil acidity.

Management

1. Practice appropriate engineering / agronomic soil conservation methods like contour bunding, contour terracing, strip cropping, relay cropping etc. Planting of erosion resistant crops (groundnut, soybean, cover crops, grasses, vetiver etc.) could reduce the constraints due to steep slope and permit cultivation.
2. Protect the soil with geotextiles / crop covers / or other artificial structures
3. Use soil binding agents to improve soil structure

2. SHALLOW SOILS

Shallow soils have a depth less than 50 cm of solum. Generally, the A horizon is very thin. These soils are found on hill tops and along the slopes. They are characterized by the presence of shallow cemented horizons. Presence of parent rocks immediately below the soil surface results in the formation of shallow soils. Soils are considered *very shallow* if they extend less than 10 inches deep before hitting an impervious layer that retards root growth. They are considered merely *shallow* if they are between 10 and 20 inches deep. Subsoils are generally not found and if present, they are very thin. These soils experience frequent droughts due to lack of sufficient soil depth and are easily subjected to erosion also. According to Gajbhiye and Mandal (2000), in India such soils are found in agroecological sub regions AER 1.1- Western Himalayas, cold arid ecoregion, 5.1 (Central Kathiawar Peninsula), 6.1 (South western

Maharashtra and North Karnataka Plateau), 6.3 (Eastern Maharashtra Plateau), 10.1 (Malwa Plateau) and 14.1 (South Kashmir and Punjab Himalayas). Majority of such soils are found in arid and semi-arid ecosystems.

Plants with surface feeding roots are generally grown in such soils. Vegetables and fruit crops like pomegranate, cherry, lemon, strawberries etc. can be successfully cultivated.

Constraints

1. Coarse soil texture and low water holding capacity
2. Limited soil depth
3. Presence of hardpan or rock or any salt deposits within 50 cm from soil surface
4. Root penetration will be restricted resulting poor crop growth
5. Due to shallowness, less volume of soil is available for roots to proliferate and hence very low nutrient availability
6. Lowering of ground water table due to over exploitation of ground water
7. Imperfect drainage due the presence of hardpans beneath soil surface
8. Subjected to severe erosion exposing the bare rock surfaces

Management

1. Growing shallow rooted crops or crops that can withstand shallowness
2. Frequent renewal of soil fertility by judicious use of soil amendments / conditioners, organic manures and fertilizers
3. Practice scientific and careful irrigation methods
4. Adopt slope protection measures on hill sides– Engineering structures / geotextiles and agronomic methods
5. Raising of grass or crops with strong anchorage

6. Cultivate erosion resistant crops
7. Deep ploughing, breaking of hardpans, application of organic manures

3. POORLY OR IMPERFECTLY DRAINED OR SLOWLY PERMEABLE SOILS / SUBMERGED SOILS

Poor drainage due to high clay content or presence of shallow water table creates imperfectly drained situations. Soils which are waterlogged or remains under water for major period are included under this group. They are usually found in backwater areas, depressions, coastal areas, river deltas etc. The natural vegetation comprises mainly moist deciduous vegetation. The major crop is rice, and in reclaimed areas other crops like coconut, banana, tubers, vegetables etc. are cultivated.

In India poorly drained soils are found in AER 13 (Hot subhumid (moist) ecoregion with alluvium derived soils) covering north-eastern Uttar Pradesh and Northern Bihar, AER15 (Hot subhumid (moist) to humid (inclusion of perhumid) ecoregion with alluvium derived soils) comprising the plains of Brahmaputra and Ganga rivers in Assam and West Bengal, AER 18 (Hot subhumid to semi arid ecoregion with coastal alluvium derived soils extending from Kanyakumari to Gangetic plains) and AER 20 (Hot humid per humid island ecoregion with red loamy and sandy soils) in Andaman and Nicobar in the east and Lakshadweep islands in the west (Gajbhiye and Mandal, 2006).

Poor drainage or slow permeability is mainly due to any of the following reasons - very high clay content; shallow water table; presence of hardpans; poor structure or poor physical conditions resulting from sodicity; landscape position; compaction; blockage of soil pores with minute colloidal particles etc. Presence of shallow water table and high capillarity create partially submerged condition with poor aeration and reduced conditions. If the surface pores are closed by fine particles, infiltration will be very low and much of the rainfall received is lost as runoff carrying fine surface soil with water in sloppy areas, in depressions it will give rise to submerged soils. Three classes of soils are found under this group.

1. **Very poorly drained:** Water is removed from the soil so slowly that free water remains at or very near the ground surface during major part of the growing season. Unless the soil is artificially drained, most of the mesophytic crops cannot be grown. These soils are frequently ponded.
2. **Poorly drained:** Water is removed so slowly that the soil is wet at shallow depths periodically during the growing season or remains wet for long periods. Free water is commonly seen at or near the surface so that most of the mesophytic crops cannot be grown, unless the soil is artificially drained. The soil, however, is not continuously wet directly below plough-depth.
3. **Somewhat poorly drained:** Water is removed slowly so that the soil is wet at a shallow depth for significant periods during the growing season. Wetness markedly restricts the growth of mesophytic crops, unless artificial drainage is provided. The soils have low saturated hydraulic conductivity / high or shallow water table or nearly continuous rainfall.

Imperfectly drained or a waterlogged soil is characterized by the presence of a thin oxidized surface layer followed by a reduced layer since the diffusion of oxygen is confined to a few millimeters of the surface soil. The sources of oxygen are that dissolved in water and that released by photosynthesizing flora. The thickness of oxidised layer depends on the supply of oxygen and its consumption by microorganisms and soil organic matter content. The oxidised surface layer is brown in colour and metabolically aerobic. It contains oxidised compounds of nitrogen, manganese, iron, sulphur etc. and their reduced counter parts are found in subsurface layer.

In submerged paddy soils the surface oxidized layer that is in direct contact with ponded water is a very thin layer. The thickness is decided by the soil organic matter content, microbial activity, soil pH and the oxygen diffusion rate. Beneath this surface layer a permanently reduced layer can be seen, the depth of which may extend up to the plough pan, usually 15-20 cm thick. This layer can undergo alternate oxidation and reduction if drainage / irrigation is provided. In the drained soils, mottles of oxidized compounds of iron and manganese (or compounds of different degrees of oxidation) can be witnessed. Beneath this, another layer of soil whose thickness depends on the depth of soil and this will be generally in oxidized condition except in low lands where the water table is very shallow.

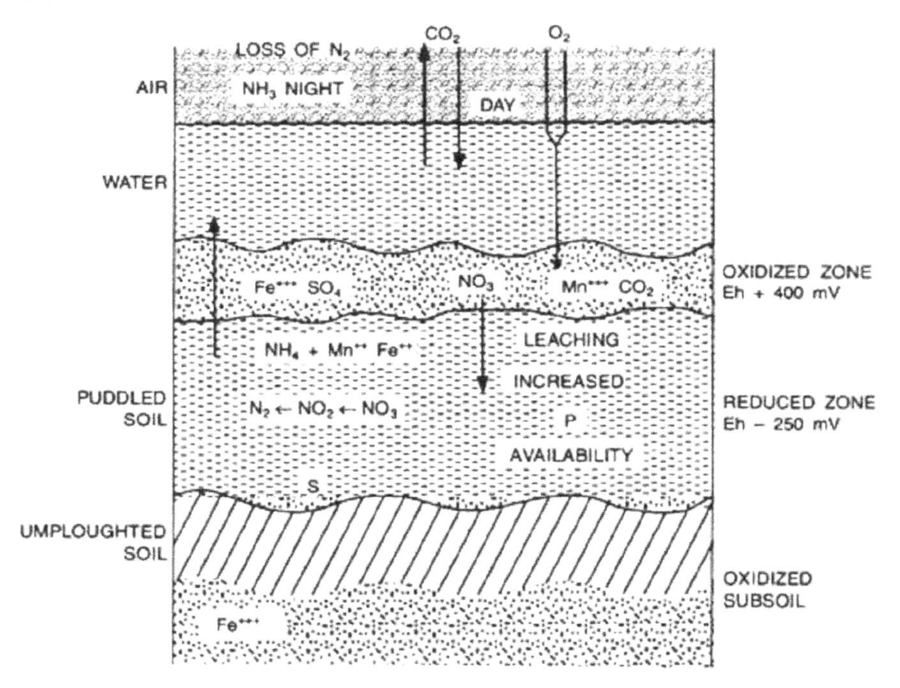

Fig. 1. Schematic diagram depicting the different layers of a submerged paddy soil

Source: http://www.soilmanagementindia.com

Characteristics

Once the soils are submerged, the oxygen from the soil pores is first depleted and water occupies that place. Simultaneously there will be a change in soil microflora. The aerobic microbes are replaced by facultative and obligate anaerobes. As a result, the following changes occur in soil due to submergence.

1. **Decrease in redox potential (Eh):** Once the soils are submerged, the air present in the soil pores is replaced by water, changing an aerobic soil to an anaerobic one and soil reduction sets in. Thus, the soils which were oxidized initially get slowly converted to a reduced soil due to acceptance of electrons. Oxidation – reduction reactions play an important role in the characterisation of submerged soils. It is a chemical reaction in which electrons are transferred from donor to an acceptor. The donor loses electrons and gets oxidized and the acceptor gets reduced. The organic matter serves as the donor of electrons and the various organic and inorganic compounds serve as the acceptors. Thus, the organic matter gets oxidized and the acceptors like nitrates, Mn^{4+}, ferric compounds etc. get reduced.

   Oxidation–reduction potential known as redox potential (Eh) is a quantitative measure of the tendency of a system to oxidize or reduce compounds. It is a measure of anaerobiosis. It is positive and high (+400 to +800 mv) in aerobic / strongly oxidized systems. On submergence, redox potential which was +ve under aerobic situation starts decreasing and gets stabilized at negative values. Since the microbes are utilizing the oxygen present in oxidized compounds like NO_3, Mn^{4+}, Fe^{3+}, SO_4^{-2}, and CO_2 they may get reduced to NO_2 / N_2, Mn^{2+}, Fe^{2+}, H_2S and CH_4 respectively, resulting a decrease in redox potential. The soil reduction follows a sequence, first nitrate being reduced followed by managanese, ferric iron, sulphates and so on. As the system gets more reduced the Eh will becomes more negative.

   Instead of Eh, the concept of pE (potential of electron) is a better choice to evaluate the redox equilibria since all these reactions are based on the electron transfer. It is the measure of electron activity. pE can be defined as the negative logarithm of electron activity and can be written as

$$pE = -\log_{10} A_{e-}$$

The pE is related to the reduction potential by the FRT factor of Nernst equation at 25 °C

$$pE = \frac{E}{0.0591} \quad 25\ ^{\circ}C$$

2. **Increase in pH of acid soils and decrease in pH of saline/alkali soils:** The increase in pH of acid soil is due the consumption of H^+ ion during the reduction of oxidized compounds in the soil. As H^+ ion in soil solution is consumed, the pH rises. The decrease in pH of saline and calcareous soils is due to release of CO_2 during organic matter decomposition,which results an increase in partial pressure of CO_2. The CO_2 reacts with water and thereby reduces soil pH. In general submergence brings the soil pH to near neutrality except for acid sulphate soils where pH will be around 5.0, even after continuous submergence.
3. **Increase in specific conductance:** The specific conductance of most of the soils increases on submergence, attains a maximum and decreases to a fairly stable value. During submergence, the pH of acid soils increases to near neutrality, where most of the nutrients are soluble and hence there will be an increase in electrical conductance. For basic soils, the pH decreases to near neutrality and this increases the solubility of most of the elements. The increase in acid soil is specifically due to release of soluble Mn^{2+}, Fe^{2+} from insoluble Mn^{4+}, Fe^{3+} oxides and hydroxides. In saline and calcareous soils, the increase is due to the dissolution of calcium and magnesium carbonates by the influence of carbon dioxide and organic acids. Higher content of organic matter favours the dissolution of calcium and magnesium carbonates by increasing the partial pressure of CO_2 on submergence.
4. **Stabilization of pH, Eh and EC:** After several weeks of submergence, various reactions in soil attain equilibrium and Eh, pH and EC get stabilized.

5. **Increase in the availability of N, P, K, Ca, Mg, Fe and Mn:** Due to increase in solubility, the availability of N, P, K, Ca, Mg, Fe and Mn increases on submergence. The availability of P increases due to solubilization of Fe/Al phosphates in acid soils and that of calcium phosphate in basic soils. The availability of Fe increases due to reduction of insoluble ferric form to soluble ferrous form. The Mn availability also increases since the oxidized Mn^{4+} get reduced to Mn^{2+} on submergence, which is the soluble form of Mn.
6. **Decrease in the availability of S, Zn and Cu:** On submergence, the sulphates get reduced to sulphides or elemental S which are not the available forms of S. Zn and Cu are soluble at a pH below 6.0 and as soil pH increases on submergence, their availability decreases. Soluble Zn and Cu can readily react with sulphides and form relatively insoluble Zn and Cu sulphides.

The nutrient availability in poorly drained or flooded soils is controlled by sequential reduction. Once O_2 in soil is depleted, facultative microbes derive O_2 from oxidised compounds and soil reduction is initiated. During this process the oxidised compounds get reduced. This reaction follows a sequence based on the potential at which each oxidised compound gets reduced and hence a sequence is followed in the reduction of oxidized compounds – called as "sequential reduction". The order is as follows.

$$O_2 > NO_3 > Mn^{4+} > Fe^{3+} > SO_4 > CO_2$$

Sequential reduction

Reaction	**Redox potential (mv)**
$O_2 + 4H + 4e \rightleftarrows 2H_2$	+400 mv
$NO_3 + H_2O + 2e^- \rightleftarrows NO_2 + 2OH$	+200 to +220 mv
$MnO_2 + 4H + 2e^- \rightleftarrows Mn^{2+} + 2H_2O$	+200 mv
$Fe(OH)_3 + e^- \rightleftarrows Fe^{2+} + 3OH$	-130 to -200 mv

$SO_4 + H_2O + 2e^- \rightleftarrows SO_3^{2-} + 2OH^-$ -120 to -230 mv

$CO_2 + 2H + 2e^- \rightleftarrows HCOOH$ -200 to -280 mv

Nutrient Transformations under Submergence

Once the soils are submerged, lot of chemical changes occurs and the nutrients are subjected to transformation. The transformations of important nutrients are discussed below.

1. Nitrogen

Nitrogen occurs in soils in different forms viz., complex organic substances, ammonia, molecular nitrogen, nitrite and nitrate. The transformations of nitrogen are largely done by microorganisms and are regulated by the physical and chemical environment of soil. The main microbiological inter-conversions are given below:

$$\begin{array}{ccccccccc} & & & & N_2 & & N_2 & & \\ & & & & \downarrow BNF & & \downarrow & & \\ \text{Proteins} & \rightleftarrows & \text{Amino acids} & \rightleftarrows & NH_4^+ & \rightleftarrows & NO_2^- & \rightleftarrows & NO_3^- \end{array}$$

Under submergence, the main processes are accumulation of ammonia, volatilization of ammonia, denitrification, nitrogen fixation and leaching losses of nitrogen. The above reactions control the availability of nitrogen and play a major role in the nutrition of rice. Most of the paddy soils are deficient in nitrogen because of the conditions that are favourable for rapid transformations and losses of nitrogen from these soils. The major transformations are described below.

Mineralization of nitrogen and accumulation of ammonia

In aerated soils, organic nitrogen undergoes mineralization to NH_4^+, oxidation of NH_4^+ to NO_2^- by the bacteria *Nitrosomonas* and oxidation of NO_2^- to NO_3^- by *Nitrobactor*. Under anaerobic situations, the mineralization proceeds only up to the NH_4^+ stage, since the absence of O_2 inhibits the activity of the *Nitrosomonas*. Thus, accumulation of ammonia in submerged soils is, therefore, a good index of the capacity of a soil to meet up the demand for nitrogen to the rice crop. Further mineralization of NH_4^+ will takes place only in oxidized layers. Nitrogen transformation in aerobic soils is as follows.

$$\text{Organic form of N} \xrightarrow[\text{Ammonification}]{\text{Mineralization}} NH_4^+ \xrightarrow{\text{Microbial oxidation}} NO_2^- \longrightarrow NO_3^-$$

Nitrogen transformation in anaerobic soil is different from that of aerobic soil and is depicted below.

$$\text{Organic form of N} \xrightarrow[\text{Ammonification}]{\text{Mineralization}} NH_4^+$$

| generally stops at this point and further oxidation only in surface oxidized layer

Nitrogen transformation occurs both in the aerobic as well as in the anaerobic layers of submerged soil. In the aerobic surface layer, nitrogen mineralization proceeds to NO_3^- form similar to well-drained soil. But this NO_3^- will diffuse downwards in response to the concentration gradient to the anaerobic layer where it gets denitrified. Thus, the presence of an aerobic layer above the anaerobic layer is the major cause of instability of nitrogen in submerged soils and results in considerable loss of nitrogen through nitrification-denitrification reactions. This process can proceed as long as NO_3^- is present in the aerobic layer, and that can be readily formed if there is NH_4^+ in the aerobic layer which get nitrified. The removal of NH_4^+ in that layer by nitrification creates a concentration gradient that causes NH_4^+ to diffuse upward from the anaerobic layer.

The overall transformations of nitrogen in submerged soils are shown in Fig. 2. In general submergence increases the nitrogen availability.

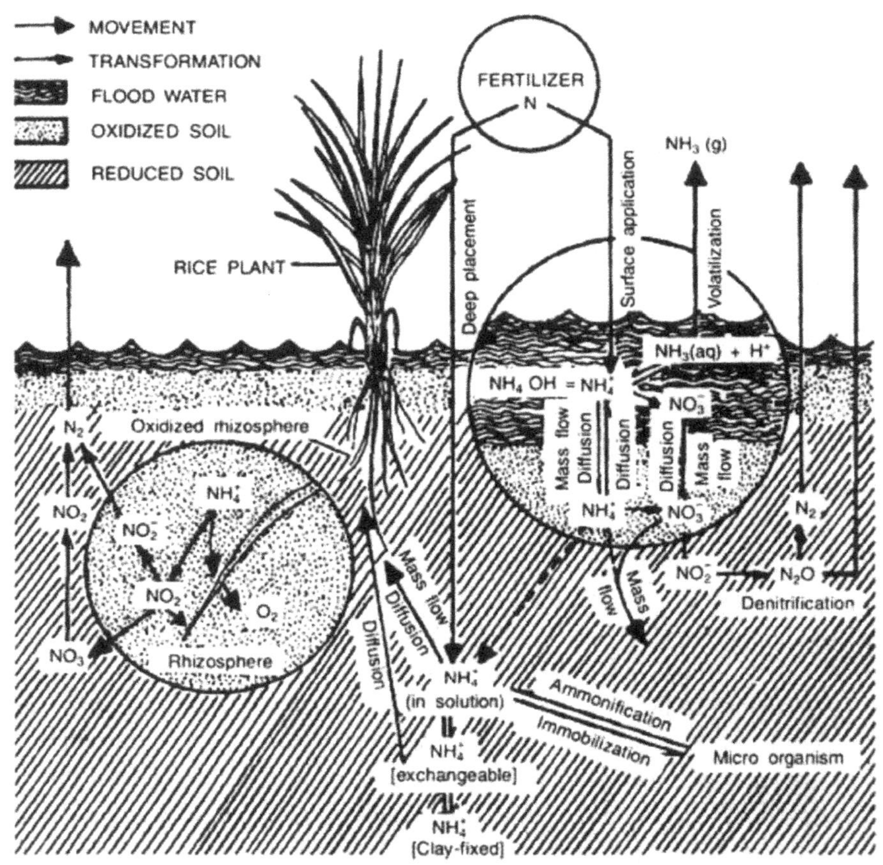

Fig. 2. Transformations of nitrogen in submerged soils

Source: http://www.soilmanagementindia.co

Phosphorus

The transformation and availability of both native and applied phosphorus is influenced by submergence though it is not directly involved in oxidation-reduction reactions in redox potential range usually encountered in submerged soils. But its reactivity with the redox elements like Fe and Al, P is getting actively involved in these reactions. On submergence, the availability of native as well as applied phosphorus increases in the soil. The phosphorus transformation in submerged soils is also associated with pH changes.

Phosphorus availability in submerged soil is linked to the behavior of iron and aluminium in acid soils and to calcium and magnesium in saline and calcareous soils. Phosphate is chemically associated with iron in two major forms in an aerobic soil ie., strengite ($FePO_4 \cdot 2H_2O$) an insoluble iron phosphate compound, and more soluble calcium and magnesium phosphates that are co-precipitated with insoluble ferric oxy-hydroxides.

On submergence of an aerobic soil, the concentration of available phosphorus initially increases and thereafter declines with the period of submergence. The magnitude of initial increase and decrease in the later period of submergence depends on the soil properties. The increase in phosphorus availability on submergence is due to the following mechanisms:

(i) Reduction of $FePO_4 \cdot 2H_2O$ to the more soluble $Fe_3(PO_4)_2 \cdot 8H_2O$. Increase in solubility of $FePO_4 \cdot 2H_2O$ and $AlPO_4 \cdot 2H_2O$ due to increase in pH coupled with the reduction of acid soils.

(ii) Release of co-precipitated or occluded phosphorus due to reduction of ferric oxy-hydroxide.

(iii) Displacement of P from ferric and aluminium phosphates by organic anions

(iv) Increase in solubility of calcium phosphates ($CaHPO_4 \cdot 2H_2O$, $Ca_4H(PO_4)_3 \cdot 3H_2O$, $Ca_{10}(PO_4)_6(OH)_2$, $Ca_{10}(PO_4)_6CO_3$ and $Ca_{10}(PO_4)_6F_2$ associated with the decrease in pH caused by the liberation of CO_2 in the calcareous soils.

(v) The release of P due to anion exchange reactions between clay and phosphate or organic anions and phosphate.

(vi) Release of P from the mineralization of organic residues – It is a very slow process and contribute small amount of soluble P in soil.

The decrease in the concentration of available P at the later period of submergence may be due to the fixation (through adsorption) of released phosphorus by clay colloids like kaolinite, montmorillonite and hydrous oxides of Fe and Al. Similar to anion exchange reactions, phosphate adsorption occurs onto variable charged minerals such as Al and Fe oxides / hydroxides and 1:1 clay minerals. Unlike the anion exchange, this reaction is not dependent on surface charge of the mineral, and it does utilize a strong covalent bond between the phosphate and a valence un-satisfied surface with no water molecule occurring between the sorbent and sorbate. Thus, the surface adsorption, also referred to as ligand exchange and surface complexation is a much stronger sorption mechanism compared to anion exchange.

In addition, the decreased concentration of phosphorus may also be due to the decreased solubility of phosphorus associated with calcium. The formation of insoluble iron oxide and hydroxides sometimes results low phosphate concentration in soil solution.

More phosphorus is released from the soil to the soil solution under submerged conditions (reduced) than that of upland soil conditions (oxidised) if the solution is initially low in phosphorus as it is found in low land rice soils. The transformations of P under alternate wetting (submergence) and drying will not be similar to that of continuous submergence. Alternate wetting and drying reduces the availability of P in Al-P fraction and that increases in Fe-P fraction.

Potassium

The transformation and availability of potassium largely depends on the nature and extent of potassium bearing minerals present in the soil. In a soil, potassium availability is further decided by two parameters. These are (i) the intensity factor (I), which is the concentration of an element in the soil solution and (ii) the capacity factor (Q), which is the ability of solid phases (soil) to replenish that element as it is depleted from solution. As plants remove K^+ ions from the soil solution, the concentration of K^+ ions in the immediate vicinity of roots is reduced and diffusion gradients are established. Potassium did not directly participate in redox reactions. However, its availability is also regulated by other cations like Fe, Mn, Al etc.

Potassium is present in soils in four forms, which are in dynamic equilibrium as follows:

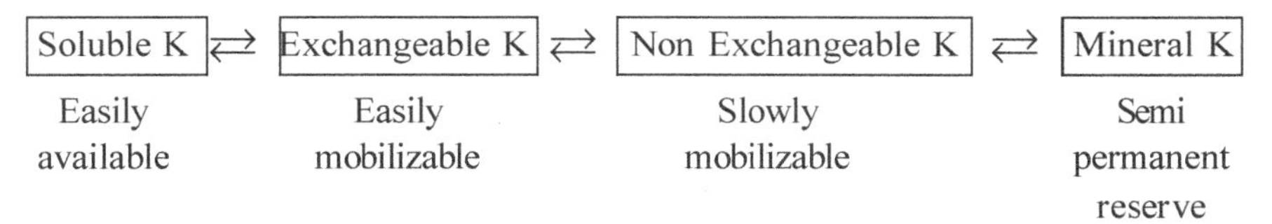

With flooding or submergence, concentration of soluble ferrous (Fe^{2+}) and manganous (Mn^{2+}) ions increases and exchangeable K^+ is then displaced into the soil solution. The increase in soluble K^+ after submergence is closely related to the ferrous ion (Fe^{2+}) content of the soil solution. In aluminium rich soils with low pH, Al also participates in these reactions similar to iron.

The release of K^+ from micas may be the contributing factor for the increase in K^+ in soil and that release depends on the various factors like tetrahedral rotation, degree of tetrahedral tilting, OH groups orientation, degree of K^+ depletion from the soil solution, hydronium ions (H_3O^+), biological activity and complexing organic acids, inorganic cations etc. Besides these, charge density and the configuration of the oxygen at exchange sites probably determines the release of K^+ and thus regulates the concentration of K^+ in the soil solution.

In general, submergence increases K availability by releasing K^+ from the non-exchangeable sites also. Continuous submergence and alternate drying and wetting increased the exchangeable potassium (K^+) content. However, contradictory reports are also there, stating that the availability of applied potassium decreases in submerged soils due to formation of Fe-K sparingly soluble complexes.

In soils dominated by 1:1 type of clays where exchangeable K is very low, submergence and subsequent increase in Fe^{++} can cause less K absorption by plant roots and plants often show the symptoms of Fe toxicity. Such environments necessitate over dose of K fertilizers in order to compete with Fe, Mn and Al and surpass Fe toxicity. Low land laterites are typical examples for this phenomenon.

Calcium and Magnesium

Calcium and magnesium also did not participate in oxidation-reduction reactions like potassium. But their availability is increased under submergence due to exchange of these cations with iron and aluminium. If water management is not practiced scientifically, these cations are subjected to leaching loss.

For rice, the calcium and magnesium requirement are generally met from liming materials like lime or dolomite. Though calcium and magnesium deficiencies are rare in lowland rice, they have been suggested as major constraints in acid sulphate soils. Since these soils were generally under-saturated with respect to Ca and Mg minerals, suggesting that mineral equilibria probably do not govern the solubility of these metals in acid sulfate soils.These soils are very rich in Fe and Al contents and the submergence enhances their solubility further. An imbalance between Ca and Mg concentration with Fe and Al may also contribute towards the wide spread Ca and Mg deficiency. Cation exchange reactions probably govern the solubility of Ca^{2+} and Mg^{2+} in acid sulfate soils (Moore and Patrick, 1989). In general, only a small amount of these elements is removed in the grain, and unless the straw is removed from the field, the total removal is small. Changes in Ca and Mg concentrations are minimum in flooded soils.

Sulphur

The main transformations of sulphur in submerged soils are the reduction of sulphate (SO_4^{2-}) to sulphides (S^{2-}) and the dissimilation of the amino acids: cysteine, cystine and methionine to H_2S. The main product of transformation of the sulphur in submerged soil is H_2S and it is derived largely from SO_4^{2-} reduction. This can cause sulphide toxicity in rice soils which are low in iron. In soils rich in Fe, sufficient quantities of ferrous Fe will be there in the soil because Fe^{3+} reduction to Fe^{2+} precedes SO_4^{2-} reduction and combine with hydrogen sulphide to form insoluble iron sulphide (FeS). In Fe deficient soils H_2S will react with other heavy metals like Zn, Cu, Cd, Pb etc. to form their insoluble sulphides, making them less available.

$$SO_4^{2-} \xrightarrow[\text{Reduction}]{\text{Submergence}} H_2S$$

$$Fe^{3+} \xrightarrow[\text{Reduction}]{\text{Submergence}} Fe^{2+}$$

$$H_2S + Fe^{2+} \longrightarrow FeS \text{ (Insoluble)}$$

The above reaction protects micro-organisms and higher plants from toxic effects of hydrogen sulphide (H_2S). In muck and sandy soils low in iron and in soils where iron is inactivated by the complex formation with organic matter, FeS formation may not take place. In that situation H_2S toxicity to the rice plant is possible.

When an acid soil is submerged the concentration of water soluble SO_4^{2-} increases initially and thereafter the concentration of the same decreases slowly. The initial increase in SO_4^{2-} concentration is due to the release of SO_4^{2-} following the increase in pH, which is strongly sorbed at low pH by clay and hydrous oxides of Fe and Al.

Rice, like other plants, absorbs sulphur primarily in the forms of SO_4^{2-} and the reduction of SO_4^{2-} to sulphide (S^{2-}) in submerged soils reduces the availability of sulphur. However, rice takes up sulphur which is oxidised as SO_4^{2-} on the root surface.

The reduction of SO_4^{2-} has three implications:

(i) Sulphur supply may become insufficient,

(ii) Zinc and copper may be immobilized and

(iii) H_2S toxicity may arise in soils low in iron.

Iron

Iron occupies most important position among the redox elements in submerged soils. The reduction of iron and the accompanying increase in its solubility is the striking chemical change that takes place when a soil is submerged. The reduction of iron is both microbiological and chemical process. The intensity of reduction depends upon time of submergence, amount of easily decomposable organic matter, active iron oxides and its crystallinity, active contents of manganese and nitrate, temperature and abundance of microorganisms. Apart from microbial reduction of ferric iron, the organic compounds produced in the course of anaerobic fermentation of organic materials dissolve insoluble ferric iron compounds to water soluble Fe complex compounds or ferric chelates.

Due to reduction of Fe^{3+} to Fe^{2+} on submergence, the colour of soil changes from brown to grey and large amounts of Fe^{2+} enter into the soil solution. It is evident that the concentration of ferrous iron (Fe^{2+}) increases initially to a peak value and thereafter decreases slowly with the period of soil submergence. Acid soils high in organic matter and active iron builds up Fe concentrations as high as 600 mg kg^{-1} within few weeks of soil submergence and then gradually decreases to 50-100 mg kg^{-1} which persists for several months. Soils high in organic matter but low in active iron content also show high concentration of Fe in soil solution but lasts for a short period only. The neutral and calcareous soils low in organic matter and active Fe, the watersoluble Fe concentration rarely exceeds 20 mg kg^{-1}. The concentration of Fe^{2+} in soil solution is influenced by redox potential, pH and partial pressure of CO_2. Higher the values of pH and Eh, the concentration of Fe^{2+} in soil solution will be lower.

Application of organic matter also enhances the rate of reduction of iron. As the partial pressure of CO_2 increases with reduction of organic matter it favours the solubilisation of Fe. Blue green algae during their growth and subsequent decomposition greatly modify the transformation and availability of iron in flooded rice soils.

On submergence, iron (Fe^{2+}) showed an initial increase due to the reduction of ferric compounds to ferrous form and reaches a peak followed by a decrease. The decrease in the concentration of Fe^{2+} following the peak is caused by the precipitation of Fe^{2+} as $FeCO_3$ in the early stages where high partial pressure of CO_2 prevails and as $Fe_3(OH)_8$ due to decrease in the partial pressure of $CO_2(pCO_2)$. Transformation of iron in submerged soils is shown in Fig 3.

Rice benefits from the increase in availability of iron but may suffer in acid soils, from an excess. It replaces cations from exchange sites and increases their availability.

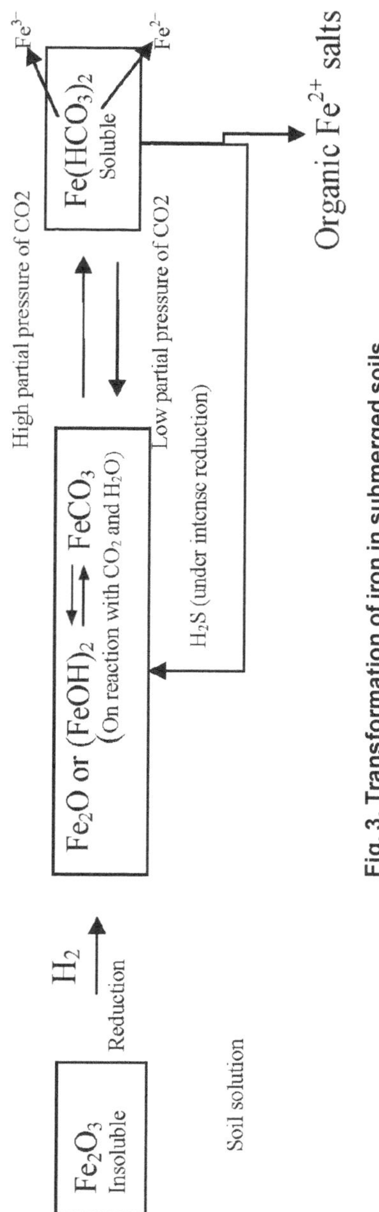

Fig. 3. Transformation of iron in submerged soils

Manganese

Manganese behaves similar to iron under submerged conditions and the transformations are governed by the redox equilibria of the system. In submerged soils manganic (Mn^{4+}) get reduced to soluble manganous (Mn^{2+}) form resulting an increase in the concentration of water soluble Mn^{2+}. The reduction of oxides of Mn concurrently takes place with the reduction of nitrates. Almost all the active Mn (water soluble + exchangeable + reducible) gets reduced within a week of flooding. The Mn concentration in soil solution is decided by soil pH, Eh, organic matter, pCO_2 and active Mn content. The concentration of Mn in soil solution represents a dynamic balance between the increased availability through reduction process, cation exchange and precipitation reactions. The concentration of Mn^{2+} increases on submergence initially, attains a peak and then decreases due to its precipitation as hydroxide resulting from increase in pH and / or carbonate resulting from decrease in pCO_2. Re-oxidation of Mn^{2+} diffusing or moving by mass flow to oxygenated interfaces in the soil is also common in submerged soils.

When an aerobic laterite soil is submerged the reduction of manganic manganese (Mn^{4+}) occurs almost concurrently with the nitrate (NO_3^-) reduction, but this reduction precedes that of Fe reduction. The concentration of Mn^{2+} (water soluble) increases initially and thereafter declines with the period of soil submergence.

$$MnO_2 + 4H^+ + 2e^- \underset{}{\overset{\text{Reduction}}{\rightleftarrows}} Mn^{2+} + 2H_2O$$

The kinetics of manganese reduction varies markedly from soil to soil. The mobilization of Mn in soil is markedly increased after submergence due to the reduction of manganic compounds to more soluble forms as a consequence of the anaerobic metabolism of soil bacteria. The transformation of Mn in submerged soils largely depends on the oxidation-reduction reactions and the reduction of Mn^{4+} occurs when the redox potential value is within a range from +200 to +400 mV.

$$Mn^{4+} \xrightarrow[\text{Submerged condition}]{\text{Reduction}} Mn^{2+} \text{ (Redox potential +200 to +400 mv)}$$

Zinc

Zinc is not involved in the oxidation-reduction process like that of iron and manganese in submerged soils. However, the reduction of hydrous oxides of iron and manganese, changes in soil pH, partial pressure of CO_2, formation of insoluble sulphides etc. influence the solubility of Zn either favorably or adversely and consequently the Zn nutrition of low land rice.

The reduction of hydrous oxides of iron and manganese, formation of organic complexing agents, and the decrease in pH of alkaline and calcareous soils on submergence are found to favour the solubility of Zn, whereas the formation of hydroxides, carbonates and sulphides may lower and ϕ, the solubility of Zn in submerged soils. Zinc deficiency in submerged rice soil is very common owing to the combined effect of increased pH, HCO_3^- and S_2^- formation.

The solubility of native forms of Zn in soil is highly pH dependent and decreases by a factor of 10^2 for each unit increase in soil pH. The activity of Zn-pH relationship has been defined as follow:

$$\text{Soil} + Zn^{2+} \rightleftarrows \text{Soil} - Zn + 2H^+$$

When an aerobic soil is submerged, the availability of native as well as applied Zn decreases and the magnitude of such decrease vary with the soil properties. The transformation of Zn in soil was found to be greatly influenced by the depth of submergence and application of organic matter.

If an acid soil is submerged, the pH of the soil will increase and thereby the availability of Zn will decrease. On the other hand, if an alkali soil is submerged, the pH of the soil will decrease and as a result the solubility of Zn will generally increase. The availability of Zn decreases due to submergence may be attributed to the following reasons.

(i) Formation of insoluble franklinite ($ZnFe_2O_4$) compound in submerged soils

$$Zn^{2+} + 2Fe^{2+} + 4H_2O \underset{\text{Reduction}}{\rightleftarrows} ZnFe_2O_{4\ (\text{Franklinite})} + 8H^+$$

(ii) Formation of very insoluble compounds of Zn as ZnS under intense reducing conditions

$$Zn^{2+} + S^{2-} \rightleftarrows ZnS_{(\text{Sphalerite})}$$

(iii) Formation of insoluble compounds of Zn as $ZnCO_3$ at the later period of soil submergence owing to high partial pressure of CO_2 (pCO_2) arising from the decomposition of organic matter

$$Zn^{2+} + CO_2(g) + H_2O \rightleftarrows ZnCO_{3\ (\text{Smithsonite})} + 2H^+$$

(iv) Formation of $Zn(OH)_2$ at a relatively higher pH which decreases the availability of Zn

$$Zn^{2+} + 2HO^- \rightleftarrows Zn(OH)_2$$

(v) Adsorption of soluble Zn^{2+} by oxide minerals eg. sesquioxides and carbonates, soil organic matter and clay minerals etc. decreases the availability of Zn. The possible mechanisms of Zn adsorption by oxide minerals are

Mechanism I: In mechanism I, Zn^{2+} adsorption occurs as bridging between two neutral sites, but in addition to this mechanism, Zn^{2+} could also be adsorbed to two positive sites or to a positive and neutral site.

Mechanism II: Mechanism II occurs at low pH and results non-specific adsorption of Zn^{2+}. In this way Zn^{2+} is retained and rendered unavailable to plants.

(vi) Formation of various other insoluble zinc compounds which decreases the availability of Zn in submerged soil eg. high phosphatic fertilizer induces the decreased availability of Zn^{2+} due to the formation of insoluble zinc phosphates.

The reversible pH change of the submerged soils, where the pH tends to increase in acid soil and decrease in alkaline soil, undoubtedly modify the Zn equilibrium concentration in the soil solution. Because the solubility of Zn minerals and Zn sobered by soil colloids is pH dependent (higher at higher pH), an increase in the pH of an acid soil when submerged will tend to decrease the Zn concentration in the soil solution.

In alkaline soil, however initially Zn uptake increases as the pH decreases after submergence. Submerged alkali or calcareous soils possess all the essential characteristics for the formation of high amount of bicarbonate (HCO_3^-) ions which makes Zn^{2+} unavailable to plants by forming insoluble $ZnCO_3$ compound.

In addition to these, the availability of Zn in submerged soil is governed by the mutual interaction of quantity **(Q)** intensity **(C)**, and kinetic parameters as regulated by the adsorption, desorption, chelation and diffusion of Zn from soils to the plant roots. The quantity-intensity relationship of Zn in submerged soil may be described by the linear form of the Langmuir type equation.

Copper

Copper is also not involved in oxidation-reduction reactions takes place in submerged / poorly aerated soils. But its mobility is affected by some of the consequences of submergence as pH, Eh, reduction of hydroxides of Fe and Mn and production of organic acids known for their complexing ability (Ponnamperuma, 1972). Flooding in general brings a decline in copper availability and this is mostly due to changes in pH since its solubility is highly pH dependent. As in the case of Zn solubility, copper also decreases 100 fold for each unit increase in pH (Fageria *et al.*, 2002).

Copper exists in soil, mainly in insoluble form and can only be extracted by strong chemical treatments which dissolve various mineral structures or solubilize organic matter. The concentration of copper in soil solutions is usually very low. At pH values below 6.9, divalent Cu^{2+} is the dominant species. Above pH 6.9, $Cu(OH)_2^0$ is the principal solution species and $CuOH^+$ at pH 7.0.

Copper exists in soils as different discrete chemical pools viz., (i) Water soluble plus exchangeable Cu, (ii) Cu associated with clay minerals, (iii) Organically bound Cu, (iv) Cu associated with different oxides in soils and (v) Residual Cu. The amount of each form of copper in soils depends on soil pH, amount of organic matter, clay content, oxides of Fe and Mn etc. All these above forms of Cu are in dynamic equilibrium in soils.

The chemical equilibria of Cu in submerged rice soils are similar to those of Zn. The mechanism for removal of Cu from soil solution in submerged soils is so pronounced that Cu is apparently removed from chelating agents that is capable of keeping the element is solution phase in upland soils.

When an acid soil is submerged, the release of copper decreases due to increase in soil pH, whereas submerging an alkali and calcareous soils, the amount of Cu in soil solution increases to a lesser degree. However, in most of the soils, in general, submergence decreases the availability of copper and thereby creates deficiency to plants, though contradictory reports are also there.

Boron and Molybdenum

Information on the effect of submergence on transformation of boron and molybdenum is meager. Since the solubility of the oxyanionic forms of these two elements is very much dependent on pH, organic matter content, clay content etc. their availability in submerged soils is highly related to the changes in soil pH. Neutral species of B (H_3BO_3) is present in soils with pH range of 5.0 to 7.0. It will dissociate only at pH above 9.2 as

$$H_3BO_3 \rightarrow H^+ + H_2BO_3^-$$

In flooded soils, where pH is around neutrality, there may not be much change in solubility of B. Ponnamperuma (1972) reported the B availability was not influenced by flooding in rice soils.The concentration of B in soil solution remains more or less constant after submergence. But there are chances for B loss by leaching and surface runoff.

Submerging an acid soil caused an increase in the amount of available Mo content during the initial periods, which remained almost unchanged at the later periods. This increase might be due to the increase in soil pH and desorption of MoO_4^{2-} from oxides and hydroxides of Fe and Mn. Pasricha *et al*., (1977) reported that available Mo in acid soils increased during the first 10 days of flooding followed by a slight decrease up to 40 days and thereafter again increased. The increase in Mo availability is mainly due to the increase pH and desorption of Mo from ferric dehydrate on soil reduction (Ponnamperuma, 1977). The availability of Mo is mainly decided by soil pH.

Constraints

1. Waterlogging : It is the major constraint that limits crop choice. Rice and other aquatic plants alone can survive under this situation. Other crops can be cultivated after providing artificial drainage and practicing suitable methods of planting like ridge and furrow or mounds.
2. Poor drainage : Most of these soils are found in coastal areas where the water table is very near to the surface making the drainage a real problem.
3. Problems with salinity / acidity: Problems with salinity is very common in coastal areas. Acid saline and acid sulphate soils are also found in the coastal belts and backwater areas due to the influence of their physiographic / geographic position and nature of parent material.
4. Elemental toxicities : Toxicities of iron, aluminium and sulphur are very common in acid soils and that of sodium in basic soils.
5. Loss of nutrients in water : Most of the nutrients are in dissolved form and are subjected to loss through drainage, run off or leaching.

6. Low organic matter decomposition : The persisting anaerobic condition causes very slow decomposition of organic matter due to low microbial activity.
7. Low microbial activity : Aerobic microorganisms are absent and only that could thrive under anaerobic condition are found in these soils which are limited in number.
8. Nutrient deficiency : Deficiency of nitrogen, sulphur, zinc and boron is quite common.

The nutrient availability under poorly drained condition is entirely different from that of a well drained soil. Certain chemical and biological changes that take place in soil on submergence have to be considered while developing management strategies.

Nutritional Management

1. **Application of organic manures:** Though many of the poorly drained soils are high in their organic matter content, application of organic manures help to reduce the toxicities and increased crop yield.
2. **Application of lime to correct pH :** Submergence itself has a self liming effect on acid soils. However, this may not be sufficient for pH correction in many soils, especially in acid sulphate soils. In such soils, liming with good quality liming materials as per lime requirement of the soil followed by washing is required for correcting soil acidity.
3. **Correction of soil salinity and sodicity:** Salt affected soils also have problems under submergence. Appropriate management practices like leaching with good quality water for saline soils, and application of gypsum for sodic and saline-sodic soils followed by leaching could correct salt problems.
4. **Use of controlled release fertilisers for N:** Under water logged conditions, nitrogen is subjected to heavy losses either in the form of nitrates or as ammonia gas. Decrease in the rate of nitrification ensures less production of nitrates and can improve N use efficiency.
5. **Improve nutrient use efficiency :** Adopt methods for increasing nutrient use efficiency (slow release / coated fertilisers, nitrification inhibitors, urease inhibitors, deep placement of urea balls etc.)

6. **Use of rock phosphate :** Easily soluble form of phosphorus results in the loss of soluble P in water and contaminate the nearby water bodies. Since rock phosphate is not water soluble it ensures slow availability in acidic conditions.
7. **Promote use of biofertilisers :** Biofertilisers are available for nitrogen, phosphorus and potassium. Their application to soils benefits the plant by regulating and ensuring slow availability of major nutrients.
8. **Scientific management of water and nutrients:** Depending on the water level, regulate the quantity, time and frequency of fertilizer application so that highest fertilizer use efficiency can be attained.

General Management / Remedial Measures

1. Select Suitable crops that can tolerate waterlogged condition
2. Select suitable planting methods like - planting on ridges or mounds
3. Use of soil amendments / conditioners: Addition of lime for acid soils; gypsum for sodic soils and organic manures for general improvement will improve soil conditions.
4. Incorporation of organics: Addition of organics namely FYM / composted coir pith @10.0 t ha^{-1} found to be optimum for the improvement of the physical properties. Application of press mud @12.5 (Latha and Janaki, 2015) improved soil physical conditions. Incorporation of paddy straw also improved soil conditions.
5. Provision of drainage: Both surface as well as subsurface drainage remove toxic materials from the field and improve aeration, ensuring better physical and chemical conditions in soil.
6. Deep ploughing: If the poor drainage is due to the presence of hardpan, break the hardpan by deep ploughing or chiseling.
7. Formation of ridges and furrows: For rainfed crops, ridges are formed along the slopes for providing adequate aeration to the root zone.
8. Formation of broad beds: To reduce the amount of water retained in black clay soils during first 8 days of rainfall, broad beds of 3-9 m

width should be formed either along the slope or across the slope with drainage furrows in between broad beds.

9. Huge quantity of sand / red soil application to change the texture
10. Contour / compartmental bunding to increase the infiltration
11. Application of soil conditioners like vermiculite to reduce runoff and erosion
12. Adopt suitable fertilizer application schedule

In India most of the waterlogged areas are located in the coastal stretches and generally face the problem of salinity. These soils are generally used for cultivation of rice or kept as swamps or marshes. The waterlogging can be seasonal and is usually associated with salinity / sodicity / permanent waterlogging and sub-surface waterlogging.

Permanently waterlogged soils are found under coastal alluvium (Tamil Nadu, Kerala, Andhra Pradesh, Maharashtra and Gujarat) and classified under Haplaqents with an area of 54403 sq. km. Deltaic alluvium (Tamil Nadu, Andhra Pradesh, Orissa and West Bengal) classified under Tropaqualfs and Quartzipsamments with an area of 87045 sq. km and alluvial soils of UP, Punjab, Bihar and Assam and peat and saline peaty soils of Kerala (2720 sq. km) are also permanently waterlogged.

4. SURFACE CRUSTED SOIL

Surface crusted soils are generally found in the soils of arid and semiarid regions. Soil crusts are specific modifications in the top soil caused by natural events such as raindrop impact and the following drying process. They are the formations of hard thin layers at the soil surface with a thickness ranging from less than 1 mm to 5 cm. When dry, these features are more compact, hard and brittle than underlying soil materials and not only decrease both the size and the number of pores, but also modify the arrangement of the pore system. The soil crusts are more compacted and packed than the underlying material. Rain impact on exposed soil is the main cause of soil crusting. Surface crusting is a dynamic process strongly related to the amount and intensity of rains and to the pedological characteristics of soils. The soils having weakly aggregated soil structure are easily broken by the impact of rain drops resulting in the formation of clay crust at the soil surface.

In temperate areas, surface crusts mainly develop on unstable loamy soils while in tropical areas, it occurs on a wider range of soils and is serious not only in the drier regions, but throughout the range of climatic regimes. At the same time, the loss of organic matter content which is very rapid in tropical areas also favours soil crusting.

Clay soils, especially those with high magnesium content and/or sodium content, are prone to soil crusting and sealing at the surface following rainfall events. This is because clay particles in soil are easily dispersible or splattered across the soil surface as rainwater hits the ground. When the water eventually recedes back into the soil, the clay is filtered onto the surface and forms a hard crust.

Mechanisms of Crust Formation

Crusting takes place mainly in the soils where the stability of surface aggregates is low. The following mechanisms are presumed to play an important role in the formation of soil crusts.

1. **Mechanical destruction of soil surface aggregates by raindrop impact:** The force with which the raindrop struck the soil surface disrupts the soil aggregates and some of the soil particles get deposited near to the original aggregates.
2. **Leaching of fine particles and their subsequent deposition in the underlying pores:** The disrupted soil particles can be easily taken away by running water or it can get deposited over soil pores and clog the pores. Once the soil pores are clogged water and air entry is prevented and initial process of soil compaction sets in.
3. **Compaction of the soil surface:** A thin film of compacted soil is formed on the soil surface which restricts both the further entry of water and the movements of fine particles in the soil pores.
4. **Cementation of the slaked soil at the soil surface due to the drying and reorientation:** Upon drying, in fact, the orientation of

the particles would contribute to the rigidity of the soil crusts. The crusts formed as a result of the above-described mechanisms of formation are called "structural crusts". Sometimes the crusts can also be formed by translocation of fine soil particles, deriving from the destruction of surface soil aggregates, and their deposition at a certain distance from their original location. In this case they are called "depositional crusts".

5. **Physical anc chemical dispersion:** Surface crusting is a characteristic phenomenon in sodic soils also, though it is not so prevalent in saline soils where Ca and Mg contents are greater which favours flocculation. In sodic soils, both physical dispersion due to rain drops impact or irrigation as well as chemical dispersion due to excess of Na in irrigation water are the main causes for crusting. Once the clay particles are dispersed, they plug the soil macropores preventing the avenues for water and roots to move through the soil. Furthermore, on drying it forms a hard cement like surface layer which restricts water infiltration and plant emergence.

Soil Properties Related to Crust Formation

The susceptibility of soils to crusting not only depends on the external factors such as raindrop impact, which acts according to the mechanisms described above, but also on the following intrinsic soil factors:

1. **Soil texture :** High clay content generally favours aggregation and reduces the rate of crust formation, although clay mineralogy and exchangeable cation composition can modify this.
2. **Organic matter content :** Organic matter is one of the most important aggregate stabilizing agents in soil. When soils are intensively cultivated, the susceptibility to crusting is increased and due to the progressive decrease of organic matter content.
3. **Sesquioxide content :** The stabilizing effect of Fe and Al hydrous oxides are commonly regarded as an important factor in aggregate formation especially in red and laterite soils.

4. **Exchangeable cations :** Saturating cations also play a leading role in crust formation due to their ability to disperse and flocculate the colloidal materials. It is well known that a high percentage of exchangeable sodium (high ESP) and in some cases exchangeable Mg, favours clay dispersion, with resulting effects on the increase of crusting.
5. **Soil water content :** Aggregates "explode" more easily when they are initially dry and then wetted suddenly. Therefore, slaking and dispersion occur more rapidly when rain falls on a dry soil, compared to soil that is already wet.

Pocket penetrometer is used to study soil crusting. It provides accurate measurements of the compressive soil strength and is expressed in "kg sq.cm^{-1} or dynes sq.cm^{-1}".

Impact of Crusting on Soil Properties

1. Prevents germination of seeds and retards root growth
2. Results in poor infiltration and accelerates surface runoff
3. Creates poor aeration in the rhizosphere
4. Affects nodule formation in leguminous crops

Sign of soil crusting is poor crop emergence. Patterns of damage depend on the levelness of the field and the pattern of soil drying across the field. Soil crusting is especially a problem in dry direct seeded fields where seed is covered by soil. It is important to address soil crusting at the time of crop establishment.

Management / Remedial Measures

1. Ploughing is to be done when the soil is at optimum moisture regime.
2. Lime or gypsum @ 2 t ha^{-1} based on the soil pH, may be uniformly spread and another ploughing is to be given for blending of amendment with the surface soil.

3. Farm yard manure or composted coir pith @ 12.5 t ha^{-1} or other organics may be applied to improve the physical properties of the soils
4. Scraping the surface soil by tooth harrow will be useful.
5. Bold grained seeds may be used for sowing on the crusted soils.
6. More number of seeds/hill may be adopted for small seeded crops.
7. Sprinkling water at periodical intervals may be done wherever possible.
8. Resistant crops like cowpea can be grown.
9. The surface crust can be easily broken by harrowing or cultivator ploughing.
10. Retaining crop residues on the surface as a protective cover.

In India crusted soils are found under AER10. Hot subhumid ecoregion with red and black soils covering parts of Malwa plateau and Bundhelkhand uplands including Baghelkhandu plateau, Narmada valley, Vindhyan scraplands and northern fringe of Maharashtra have the crusted soils. Here the black soils are interspersed with patches of red soils. The soil belongs to soil orders Entisols, Alfisols, Inceptisols and Vertisols. Rainfed agriculture is practiced here with crops like rice, millets, pigeon pea, green gram and black gram in Kharif season. In areas where irrigation facility is available, rice and wheat are cultivated during rabi season.

5. SOILS WITH SUBSOIL HARDPAN

Hardpan is a dense layer of soil, usually found below the uppermost top soil layer. There are different types of hardpans, all sharing the general characteristic of being a distinct soil layer that is largely impervious to water. Some hardpans are formed by deposits in the soil that fuse and bind the soil particles. These deposits can range from dissolved silica to matrices formed from iron oxides and calcium carbonate. Others are man-made, such as hardpan formed by compaction from repeated ploughing, particularly with moldboard ploughs, or by heavy traffic or pollution. Subsoil hardpan is commonly found in red soils. Though soil is fertile, crops cannot absorb nutrients from the soil which leads to reduction in crop yields. The reasons for the formation of sub surface hardpan in red soils is due to the illuviation of clay to the subsoil horizons coupled with cementing action of oxides of iron, aluminium and calcium carbonate.

Mechanism of Hardpan Formation

Soil structure and soil pH are highly related to hardpan formation. Acid soils are most often affected due to the propensity of certain mineral salts, most notably iron and calcium, to form hard complexes with soil particles under acid conditions.

Another major determinant is the soil particle size. Clay particles being minute in size, it occupies the inter space and restricts the passage of water, reducing infiltration and hence drainage. Soils with high clay content are also easily compacted and affected by man-made discharges. Clay particles have a strong negative electrostatic charge and will readily bond to positively charged ions dissolved in the soil-water matrix. Common salts such as sodium ions contained in wastewater can fulfill this role and lead to a localized hardpan in some soil types. This is a common cause of septic system failure due to the prevention of proper drainage in field. Presence of high clay content / $CaSO_4$ / $CaCO_3$ / $CaSO_4$ / iron and aluminium oxides etc. are responsible for the formation of hardpans.

Laterite soils generally have hardpans of iron oxides while in arid tracts, the hardpans are mainly of carbonates and sulphates of calcium and magnesium. The position of the hardpans directly affects plant growth by restricting root advancement to deeper layers of soil, thereby limiting the nutrient extraction.

Constraints

1. The subsoil hardpan is characterized by high bulk density (>1.8 Mg m^{-3}) which in turn lowers infiltration, water holding capacity, available

water and movement of air and nutrients with concomitant effect on the yield of crops.

2. Hardpans present beneath surface soils restricts the downward water movement and results impeded drainage leading to water stagnation or loss through run off.
3. It restricts root growth and thereby plant growth also.

Management / Remedial Measures

1. The immediate solution is breaking of the hardpan either mechanically or through the use of soil amendments. The broad fork is a manual tool specifically designed for this task; a digging fork or a spade might also be used. The chisel plough does a similar job with the help of a tractor. The soil amendments added will alter the soil structure and promote the dissolution of the hardpan.

   Chiseling technology : The field is to be ploughed with chisel plough attached to a tractor. The heavy iron plough has to be drawn at 50 cm interval in both the directions. Chiseling helps to break the hardpan in the subsoil besides it ploughs up to 45 cm depth. Farm yard manure or press mud or composted coir pith @ 12.5 t ha^{-1} is to be spread evenly on the surface. The field should be ploughed with country plough twice for incorporating the added manures. The broken hardpan and incorporation of manures make the soil to conserve more moisture
2. Addition of soil organic matter through manure, compost or peat can prevent hardpan formation
3. Improve local drainage and promote the proliferation of earth worms that can, over time, break relatively thin hardpan layers.
4. More difficult hardpans may be further improved through the action of both adjusting the soil pH with lime if the soil is acidic, and otherwise with the addition of gypsum. However, unlike when employing mechanical means, breaking a hardpan through the use of amendments may require several years, and even then, success is not assured.

5. Raising vegetative barriers of vetiver or lemon grass across the slope and along the contours at 0.5 m vertical interval will ensure better *in situ* moisture conservation in drylands of Vertisols and can prevent hardpan formation.

6. HIGHLY PERMEABLE SOILS

This group include sandy soils containing $>$ 70 per cent sand and clay content $<$ 18 per cent. Generally, occurs in coastal areas, river deltas and in the desert belts. From these soils, water is removed rapidly. Internal free water occurrence is commonly very rare or very shallow. The soils are commonly coarse-textured and have high saturated hydraulic conductivity.

In India, highly permeable soils are found in the coastal areas and in the arid and semi-arid tracts of AER 2. These soils are widely distributed in Rajasthan, Gujarat, Haryana and Punjab. All the States with coastal belts have highly permeable soils.

Constraints

1. Very high infiltration rate and hydraulic conductivity: Since these soils are very loose and poorly structured, they have large number of macropores which allows water to drain easily. Hence infiltration rate and hydraulic conductivity are very high.
2. Low moisture retention and water holding capacity: Presence of few micropores and low organic matter and clay content reduce the water retention of these soils to a minimum.
3. Generally poor in organic matter and nutrients
4. Low in bases like Ca, Mg and K
5. Low in micronutrients viz., Zn, Cu and B
6. Poor structure or weak aggregation, hence applied nutrients are easily lost from soil

7. Less anchorage to the crops grown
8. In extreme hot areas, root growth is limited due to high soil temperature

Management / Remedial Measures

1. Compaction of soils: Providing compaction to sandy soils by rolling 400 kg stone roller 10 times at optimum moisture will help to conserve soil moisture. After a shallow ploughing, crops can be raised.
2. Apply organic materials like farm yard manure, compost, press mud, sugar factory slurry, composted coir pith, sewage sludge etc. to improve soil structure.
3. Crop rotation with green manure crops like sunhemp, sesbania, daincha, kolinchi etc. can be practised.
4. Frequent irrigation with less quantity of water should be provided.
5. Provide mulching using natural or artificial materials.
6. Frequent split application of fertilizers and use of slow-release fertilizers like neem coated urea should be advocated.
7. Addition of tank silt for coastal sandy soils is recommended for enhancing their productivity. Application of tank silt or black soil at the rate of 25 t ha^{-1} per year along with FYM, composted coir pith or press mud @ 25 t ha^{-1} is found to be good for highly permeable red soils. Deep ploughing the field with moldboard plough or disc plough during summer also improved the water holding capacity of the soil.
8. Providing asphalt sheet / polythene sheet below the soil surface to reduce the infiltration
9. Application of clay soil up to a level of 100 t ha^{-1} based on the severity of the problem and availability of clay materials is a good option but practically difficult.

7. HEAVY CLAY SOILS / HEAVY CRACKING SOILS (VERTISOLS)

Clay soils are referred as heavy soils. These are also known as slowly permeable soils. Heavy clays have very high water-holding capacity, but most of the water is tightly bound and not available to plants. The humus content is often higher compared to other mineral soils. They form cracks when they are dry. These soils have a very good ability to improve their structure through freezing/thawing and drying/wetting. In cold winters the clay freezes apart and forms a very favourable aggregated structure in the topsoil layer. If the clay dries out without having been frozen, it can become very stiff and difficult to work. In the water-saturated state these soils are sticky and very impermeable to water. Due to the high clay content, the nutrient content is very high. Tilling these soils in wet conditions leads to soil compaction. Normally these soils come under soil order Vertisols. The characteristics are:

1. Black or dark coloured soils with high clay content (>30%), especially rich in montmorillonite clay. The bulk density is also high (1.5 to 1.8 mg m^{-3})
2. They swell on absorption of water and shrink on drying. During this process, soils develop cracks about 1 cm wide. The cracks are wedge shaped and on drying, soil will fall into these cracks resulting inversion of soil. Hence there is no horizonation. They do not exhibit any pedogenic process like eluviation or illuviation because of the inversion.
3. The high clay content of swell-shrink nature reduces the permeability and hence these soils are poorly drained.
4. They are extremely hard when dry and highly sticky and plastic when wet making the cultivation operation much difficult.
5. The nutrient and water holding capacity is high, but water availability is poor due the dominance of smectite clay which held water tenaciously.

6. Under poor drainage, leaching of weathering products is restricted and hence pH > 7.0 is shown by these soils. Calcareous nature of the soil results a pH range of 7.8 to 8.7 and sodic conditions if persists it may go up to 9.5.
7. Being rich in 2:1 clay minerals, they have very high CEC (30-60 cmol kg^{-1}) and base saturation with high contents of Ca and Mg. Calcium carbonate nodules are present
8. The fertility status is high, but unbalanced. N and P are generally low and high in Ca and Mg.
9. Deficient in micronutrients. Wide spread deficiency of Fe, Mn, Zn and B are reported from these soils.

Repeated cycles of shrinkage and swelling results opening and closing of cracks, causes a sort of self- mulching. In the long-term, soil material falling in cracks during each dry season produces the mixture of the material and the upper horizon can be very deep due to the continuous internal turnover. The process of opening and closing of cracks produces characteristic repeated mounds and depressions on the shrink-swell and cracking clays. These microreliefs are known as **gilgai.** In agricultural soils, gilgai may be difficult to observe due to tillage.

The contraction of the aggregates during the dry season results the formation of deep wedge-shaped cracks (spenoids) at the surface extending deep to subsoils. The cracks get filled with surface soil by the wind action or animal activity. During subsequent rainy season they get swelled. Since there is no more space available for expansion since the cracks are already filled, they push upwards and/in horizontal plane in an ellipsoidal way, glide passing each other, resulting in the formation of smooth polished ped surfaces called **slickensides** in subsoil. Continuous expansion / contraction leads to a uniform distribution of clay particles in the surface of aggregates. These pressurized reflective surfaces ie., **slickensides** is a characteristic phenomenon of Vertisols.

There are two broad groups of Vertisols

1. **Self-mulching Vertisols:** These have a fine (granular or crumb) surface soil structure during the dry season, produced by desiccation

and shrinkage. When such soils are ploughed, the clods, after being subjected to repeat wetting and drying, disintegrate to produce fine crumb or granules.

2. **Crusty Vertisols:** These have a thin, hard crust in the dry season. When ploughed, these soils produce large, hard clods that may persist for 2 to 3 years before they get crumbled enough to permit the preparation of a good seedbed. Such soils require mechanical tillage if they are to be cultivated.

Constraints

1. Very difficult to maintain optimum moisture content: Dominance of expanding 2:1 clays makes the soil to either swell or shrink depending on the moisture availability. High clay content allows water to escape to atmosphere very slowly and hence drying will take place at a slow pace. Similarly, the swelled soils cannot be drained easily. These two properties make the water management very difficult in these soils.
2. Heavy soils have very hard consistence when dry and very plastic and sticky ("heavy") when wet. Therefore, the workability of the soil is often limited to very short periods of medium (optimal) water status. However, tillage operations can be performed in the dry season with heavy machinery. Mechanical tillage in the wet season causes serious soil compaction. Ploughing should be done at optimum moisture content of soil.
3. These soils are imperfectly to poorly drained and hence leaching of soluble weathering products is limited. This is due to the very low hydraulic conductivity. Once the soil has reached its field capacity, practically no water movement occurs. In high rainfall areas such soils have to face frequent flooding causing much havoc to domestic life as well as to agriculture.
4. The structural stability of heavy clay soils remains low. They are therefore very susceptible to water erosion.
5. Most of the heavy clay soils belonging to Vertisols are chemically rich and are capable of sustaining continuous cropping. They do not

necessarily require a rest period for recovery; since pedoturbation continuously brings subsoil to the surface. Pedoturbation is a process of soil mixing due to the fall of surface soils to the cracks and allows soil mixing leading to homogenization of solum to varying degrees. However, the overall productivity normally remains low, especially where no irrigation water is available. N, P, S, Fe, Mn and Zn are normally deficient.

Management / Remedial Measures

1. Follow scientific tillage and water management: Practice tillage operations only under optimum moisture conditions.
2. Improve surface drainage in poorly drained soils by adding materials like sawdust or coir pith and /or provision of subsurface drainage / addition of soil amendments.
3. Adopt water harvesting techniques in dry areas.
4. Application coir pith / organic manure / green manure improves physical condition of soil.
5. Slopes above 5 % should not be used for arable cropping, and on gentle slopes contour cultivation with a groundcover crop is advisable. While terracing, sufficient surface drainage must be provided.
6. Planting of seedlings can be done in pits (preferably of size $1m^3$) filled with washed coir pith.
7. Follow scientific irrigation and nutrient application. Fertigation is an ideal option to these soils.
8. Select suitable cropping systems.
9. Make raised beds to assist drainage and to reduce trampling of the soil.

Heavy cracking soils (Vertisols) are found in AER 5, 6, 7, 8 and 10 of India covering Central Highlands, Gujarat Plains, Kathiawar Peninsula and Decan Plateau. The famous black cotton soils belong to Vertisols. Imperfect drainage is one of the major problems in these areas which limits optimum root ramification and oxygen availability especially in low lying areas. The areas near to the sea face salinity problems also. Presence of cracking clay soils having narrow workable moisture conditions make the tillage /cultivation activities very difficult. During rainy seasons, heavy runoff occurs resulting in huge loss of soil.

Traditional rainfed agriculture and irrigated agriculture are common. In hot semi-arid ecoregions with black soils, dry land farming is practiced. Here sorghum, pearl millet, pigeon pea, groundnut, maize, soybean, pulses etc. are cultivated during kharif season and safflower, sunflower and grams during rabi season.

8. FLUFFY PADDY SOILS

They are characterized by low bulk density of top soil resulting sinking of farm animals and labourers as well as poor anchorage to rice seedlings.The traditional method of preparing the soil for transplanting rice consists of puddling, which substantially breaks soil aggregates into a uniform structure-less mass. Under continuous flooding and submergence of soil for rice cultivation in a cropping sequence of rice-rice-rice, the soil particles are always in a state of flux and the mechanical strength is lost leading to the fluffiness of soils. In Kerala, fluffy paddy soils are prevalent in Chittoor district of Palakkad (*Poonthalpadam* soils, *Poonthal* is the local name for sinking). It is formed as a result of impossible drainage and subsequent dispersion. A continuous rice-rice cropping sequence is practiced in these soils.These soils are rich in minerals and had a very high CEC and base saturation. Soils are deficient in organic matter, available N, P and Zn. In several places, the sodium content is also high and soils are in a highly dispersed state.

Constraints

1. Mechanical cultivation is not possible in fluffy soils. Sinking of draught animals and labourers is one of the problems during puddling in rice fields which is an invisible drain of finance for the farmers due to high pulling power needed for the bullocks and slow movement of labourers during the puddling operations.

2. The low bulk density and fluffy nature of soil adversely affects anchorage to the roots and the potential yield of crops declines.
3. Impeded drainage and poor physical properties are the major problems in this area.
4. They are also low in organic matter content, available N, P and Zn.

Management / Remedial Measures

Following practices are adopted to overcome the problems.

1. The irrigation should be stopped 10 days before the harvest of rice crop.
2. After the harvest of rice, when the soil is under semi-dry condition, compact the field by passing 400 kg stone roller or tar drum filled with 400 kg of sand, 8 times.
3. The usual preparatory cultivation is carried out after compaction.
4. Addition of plant residues is very important for regeneration and maintenance of soil structure in the transplanted rice ecosystem, but for various reasons, the amount of residue being returned to the soil is inadequate. The added organic matter will improve soil physical health, and helps nutrient retention, improve soil biological activity and improve crop production.
5. If better drainage can be provided in fluffy soils, several problems can be solved.

9. ORGANIC SOILS

Soils that are rich in organic matter and that contain plant materials under different grades of decomposition are categorized under organic soils and classified under soil order Histosols as per soil taxonomy. Histosols are soils that are composed mainly of organic materials. They are defined as having 40 cm or more of organic soil material in the upper 80 cm and have an organic carbon content of 12 to 18 per cent, or more, depending on the clay content of the soil. Such soils are formed under restricted drainage or anaerobic flooded conditions where organic matter is formed / accumulated at a faster rate than its destruction due to a decrease in microbial decay rates. This happens mainly in extremely wet areas or under water; thus, most of these soils are saturated year-round. They are found in the low-lying marshy areas or to the soils confined to the depressions formed by drying of lakes.

Globally such soils occupy an area of 12 M ha of South East Asia, West Africa and along the north-eastern coast of South America. In India, they occur in localized pockets in States of Kerala, Orissa, West Bengal (Sundarbans), Goa and south eastern Tamil Nadu where they are mostly associated with mangrove forests. Organic soils are used extensively for vegetable production and with proper management it will produce excellent crops year after year. Organic soils contribute less than one per cent of the total land area of world.

Organic soils are usually differentiated on the basis of the state of their decomposition. Deposits which are slightly decayed are called peat soils where it is easy to identify the origin of plant materials. Completely decayed materials where it is virtually impossible to identify any of the original materials are called muck soils. Peat soils can be fine or coarse textured depending on the nature of the deposit. Well-decomposed mucks are very fine and when dry, are quite powdery and subject to wind erosion.

Marshes, bogs and swamps provide conditions suitable for the accumulation of organic soils. This environment encourages the growth of many plants and trees. Numberless generations, over thousands of years grow and die and sink down into the water. The water shuts out the air and prohibits rapid oxidation and acts as a preservative. The breakdown is brought about by fungi and anaerobic bacteria which aids in the synthesis of humus. As one generation of plants follows another, different layers of organic residues are created. In some areas, different layers accumulate in profiles. Deep water plants may be succeeded by weeds and sedges, followed by mosses and shrubs and finally hardwoods or deciduous trees. In other areas, fluctuating water levels and climatic events have altered the accumulation and changed the sequence of the profiles. These soils are dominantly observed in coastal inter-tropical lowlands.

These soils are dark to almost black in colour with abundant organic matter (20-40%) content and fine in texture. Organic soils have very low bulk density and are poorly drained because the organic matter holds water very well. They are strongly acidic (pH 3.5-4.0) and majority of soils are very deficient in major plant nutrients, which were washed away from the consistently moist soil. Such soils show accumulation of ferrous and aluminium sulphates and iron pyrite especially in tidal swamp areas. These soils are generally very difficult to cultivate because of the poor drainage and often low chemical fertility. On drying the soils show remarkable tendency for subsidence. There is a great risk of the organic matter becoming dry powder and eroding under the influence of drying winds. A tendency towards shrinkage and subsidence is common. These soils have the problem of deficiency of micronutrients such as boron, copper and zinc. The organic soils that are acid sulphate in nature have the characteristics and constraints as that of acid sulphate soils.

Constraints

1. Subsidence: Organic soils have a major chronic problem of subsidence where the soils subside at a steady rate and permanently lower the surface elevation of the soil. Major factors responsible for subsidence are oxidation of the soil organic matter, soil shrinkage, wind erosion, water erosion, and height of the water table. The rate of subsidence varies, depending on the frequency of wind erosion, the organic-matter content of the soil, the degree of water-level control, and the methods of cultivation.
2. Difficulty in operating the agro-machinery: Since the soils are not steady and stable use of agro-machinery is very difficult. Cultivation operations like weeding, spraying and harvesting operations are hampered.

3. Problems due to salinity: Since most of these soils are formed near sea or lake, problems associated with salt accumulation are common. High concentrations of soluble salts in soils can prevent or delay germination of seeds and can seriously damage established plants. Salt problems have become common especially during drought periods.
4. Erosion hazards: These soils are subjected to severe erosion by both wind and water, the wind erosion being more severe in dry areas because of very low bulk density of particles.
5. Nutrient availability: Very low pH affects nutrient availability. Deficiency of P and micronutrients like Zn and Cu is widely observed.
6. Problems due to acid sulphate nature: Organic soils of acid sulphate nature show all the constraints associated with high acidity, iron, aluminium and sulphur toxicity.

Reclamation / Management

1. Provide shallow drainage channels or ditches of 1 m deep at 20-40 m intervals.
2. Controlled burning of the peat to free nutrients and to raise the pH of the surface soil.
3. Because of the high organic carbon content, organic soils always show Cu deficiency. Copper applied as fertilizer slows down the activity of enzymes and reduce subsidence by about 50%. It is recommended that 14 kg of Cu ha^{-1} be applied for the first three years of initial cultivation followed by 5 kg Cu ha^{-1} every second year, particularly when onions, carrots or lettuce are grown.
4. The level of the water table influences crop production and has a major effect on the rate of subsidence. Therefore, the water table should be maintained at a level which will keep subsidence to a minimum and at the same time produce optimum crop yields.
5. Adopt erosion control measures
6. When the pH is 5.1 or lower, application of lime is generally recommended, particularly if the soils are deficient in calcium.

7. If irrigation is needed in areas with salt problem, provide overhead irrigation with good quality water.
8. Fertilise organic soils with nutrients as per soil test data.
9. For soils that are acid sulphate in nature follow the reclamation practices that will improve soil pH and reduce the toxicity of Fe and Al.

In India organic soils are found in Jammu and Kashmir and Himachal Pradesh. In Kerala and West Bengal, organic rich soils are found in lower topographic positions in a permanent water saturated environment. High acidity and high content of sulphidic material are associated with these soils. These soils present numerous nutrient management problems and need special management for optimum returns.

CHAPTER

SOILS WITH CHEMICAL PROBLEMS

4

1. ACID SOILS

About 30 % of Indian soils are acidic in nature, covering 49 million ha of land. Acid soils are generally found in high rainfall areas. In India, they are distributed mainly in Assam, West Bengal, Meghalaya, Himalayan belt, Andhra Pradesh, Telangana, Karnataka and Kerala. Acid saline soils are present in Kerala (*Kuttanad, Pokkali* and *Kaipad*), Odisha and West Bengal (Sundarbans area).

Acid soils are characterized by the preponderance of H^+ ions. The H^+ ions are held on the clay particles or on the organic matter. H^+ found in soil solution, Fe^{3+} and Al^{3+} found in soil matrix or in solution are largely responsible for soil acidity.

Sources of Soil Acidity

1. **Climate:** In high rain fall areas, most of the bases like Ca, Mg, K, Na etc. will be leached out, resulting an accumulation of Fe and Al in soil which are acidic in nature. Furthermore, the carbonic acid produced may displace the remaining bases present in soil and H^+ ions occupy their place making the soils more acidic.
2. **Parent materials:** Soils derived from acidic parent materials are acidic in reaction.

3. **Humus and organic matter:** Several organic acids are formed during the organic matter decomposition. They behave like weak acids and dissociate H^+.

4. **Aluminosilicate clays:** The H^+ ions due to permanent charges (by isomorphous substitution) and pH dependent charges (arise from the structural OH groups on corners and edges of soil clay minerals) may dissociate in to H^+ ions in slightly acidic to alkaline condition. Many such aluminosilicate clay minerals are coated by amorphous Al and Fe oxides. When pH increases to 5.0, such coating may get dissolved and unblock the exchange sites. Further hydrolysis may again release H^+ ion from the above clay minerals.

   When the pH is below 4.5, H^+ ions replace some of the Al^{3+} ions and release to soil solution. The Al^{3+} undergoes subsequent hydrolysis. In acid soil below pH 5.0, Fe also behaves similarly.

$$Al^{3+} + H_2O \rightleftarrows Al(OH) + H^+$$

$$Al(OH) + H_2O \rightleftarrows Al(OH)_2 + H^+$$

$$Al(OH)_2 + H_2O \rightleftarrows Al(OH)_3 + H^+$$

Gibbsite gets precipitated in acid range

$$Al(OH)_3 + H_2O \rightleftarrows Al(OH)_4 + H^+ \text{ (Occurs in alkaline range)}$$

5. **Oxides, hydrous oxides and polymers of Fe and Al:** They undergo stepwise hydrolysis and release H^+ ions

6. **Production of acidic substances:** In acid sulphate soils which are typical peat saline soils, the acidity is due to production of suphuric acid on oxidation of ferric sulphate (jarosite).

7. **Contribution from environment :** When the electric discharge occurs in the atmosphere during rainy season, atmospheric nitrogen, sulphur and oxygen combines to form sulfuric / nitric acids. In industrial area coal is used as energy sources which produce SO_2 in the environment, this SO_2 react with water (H_2O) in the atmosphere and bring back to the surface as acid rains.

$$N_2 + 2O_2 \rightarrow 2NO_2$$
$$2NO_2 + H_2O \rightarrow HNO_3 + HNO_2$$
$$S + O_2 \rightarrow SO_2$$
$$SO_2 + \frac{1}{2}O_2 + H_2O \rightarrow H_2SO_4$$

8. **Production of CO_2:** During organic matter decomposition, CO_2 is produced and get dissolved in water forming carbonic acid, which may dissociate and release H^+ ions
9. **Nature of soil colloids:** When H^+ ion is the dominant cation of soil colloids, the soil becomes acidic.
10. **Low Percentage Base Saturation (PBS) and kinds of bases:** Soils with low PBS (less than 50) will be acidic in reaction. If the dominant cation is Na, the pH will be high compared to Ca or Mg.
11. **Soil management:** Addition of acid forming fertilisers results soil acidity.
12. **Oxidation reduction potential:** On submergence, the soil gets devoid of oxygen and anaerobic microorganisms utilize the oxidized compounds present in the soil and soil reduction occurs. During this process, the H^+ ions are consumed, causing an increase in soil pH.On drying/draining of the soil the reverse process occurs and soil acidity increases.

Other sources of soil acidity include use of S containing pesticides, addition of acidic effluents from factories, release of mine wastes etc.

Characterization of Soil Acidity

Soil acidity is classified in to three types based on the source of H^+ ion.

1. Active acidity
2. Exchangeable or salt replaceable acidity
3. Potential acidity

1. **Active acidity:** It is a measure of H^+ ion activity in soil solution. The Al^{3+} present in the soil solution undergoes hydrolysis and releases H^+ ions, contributing to active acidity. The active pool of H^+ ions is in equilibrium with the exchangeable H^+ions that are held on cation exchange complex. It is very small compared to other two acidities. But this will directly affect the plant roots and microbes present in the rhizosphere. By measuring pH of a soil suspension, active acidity is measured. The following equation express the relation.

$$HA \rightarrow H^+ + A^-$$

H^+ = Active acidity

HA = Potential acidity

2. **Exchangeable acidity:** It is mainly associated with exchangeable H^+ and Al^{3+} adsorbed on soil colloids. Exchangeable acidity is more evident in strongly acidic soils while its contribution is meager in moderately acidic soils. It is mainly represented by H^+ and Al^{3+} and to a lesser extent by Fe^{3+} and Mn^{4+} that are easily exchangeable by other cations in a simple unbuffered salt solution. It is the acidity extracted by 1 M KCl solution and determined by titrating the extract with standard alkali.

3. **Potential acidity:** It is also known as *extractable acidity* or *reserve acidity* or *residual acidity* or even *titratable acidity* and in some times confusingly referred as *exchange acidity*. It is mainly associated with non-exchangeable H^+ and Al^{3+} that are bound in non-exchangeable sites by organic colloids and silicate clays. It is that what remains in soil after the neutralization of active acidity and exchangeable acidity. It indicates the acidity that can be neutralized by liming materials, but cannot be detected by salt replaceable techniques. It is several times greater than active acidity and exchangeable acidity. It represents the acidity extracted by $BaCl_2$-TEA buffer solution at pH 8.2 (Blakemore *et al.*, 1987).

Buffering Capacity of Soil

It is the resistance towards the change in pH of a soil. It is defined as the tolerance to changes in pH of a soil solution. The soil will behave as a weak acid and resist changes in pH. Under acidic condition, the H^+ ions in soil solution will be in equilibrium with H^+ and Al^{3+} ions in the exchange complex. When liming materials are added to soil, first the H^+ ions in soil solution gets neutralized. Then the Fe^{+3} and Al^{+3} in the exchange complex undergo hydrolysis and release H^+ ions. This will continue as long as sufficient quantities of liming materials are added to neutralize exchangeable acidity and residual acidities, resulting a pH change. The buffering capacity will be high if CEC is high. The buffering capacity will be higher for soils high in clay and organic carbon.

The degree of buffering capacity is highest between pH 4.5 and 6.0 and drops below pH 4.5 and above 6.0. The buffering action is due to the influence of weak acids and their salts. Carbonates, bicarbonates and phosphates act as buffering agents. Another soil constituent that affects buffering capacity is organic acids produced in the soil; they are weak acids and serve as excellent buffering agents. Soil colloids both organic and inorganic also had profound influence on buffering capacity of soil. The buffering capacity of soil protects soil from sudden changes in soil pH and protects crops. It gives information on the quantity of amendments required for reclamation.

Problems Encountered in Acid Soils

Under extremely acid conditions, the plant growth will be restricted. The chemical changes consequent to lowering of soil pH can restrict the availability of essential plant nutrients and increase the availability of toxic elements. Most of the problems are either due to the direct or indirect effects of acidity. Problematic acid soils have a pH of less than 5.6 and usually below pH 5.0. The low soil pH is associated with a number of soil chemical and biological characteristics that manifest themselves as the components of the problem acid soil syndrome. These components may adversely affect plant growth. To establish wheather a problem related to acidity exists, first soil pH should be measured and if the pH is low enough, more detailed soil / plant test should be undertaken.

Soil pH will influence both the availability of soil nutrients to plants and how the nutrients react with each other. At a low pH many elements become less available to plants, while others such as iron, aluminum and manganese become toxic to plants and in addition, aluminum, iron and phosphorus combine to form insoluble compounds. In contrast, at high pH levels calcium ties up phosphorus, making it unavailable to plants, and molybdenum becomes toxic in some soils. Boron may also be toxic at high pH levels in some soils. The common problems encountered are described below.

Toxic Effects

A. **Direct toxicity:** Large quantities of H^+ ions present in the soil is toxic to crop roots and cause root decay which itself is a severe problem since without proper root growth plant surveillance is not possible. Soil acidity reduces the cell elongation because it adversely affects the development of meristematic tissues of root tip and restrict root growth.

B. **Toxicities due to nutrients**

1. **Iron toxicity:** It is a serious problem in paddy growing wetland acid soils. pH below 5.0 under wet conditions, Fe is present in soluble ferrous form. Though submergence will increase soil pH, most of the paddy soils face the problem of Fe toxicity due to excess amount of ferrous iron in soil in response to soil reduction. Plant roots will take Fe^{2+} and brown coloured rusty spots are seen on the leaves. This is known as **bronzing.**
2. **Aluminium toxicity:** Common in soils below pH of 4.5. Al retards root growth and cell multiplication. Root decay is the common symptom.

3. **Manganese toxicity:** At pH 4.6 and below, toxicity can be a continuing problem where high soil Mn^{2+} is primarily caused by low pH. Soil manganese exists in an equilibrium between plant available manganous manganese (Mn^{2+}) and unavailable manganic manganese (Mn^{3+}). Plants suffer from toxicity when they absorb too much Mn^{2+}. The balance between available and unavailable forms of manganese is influenced by chemical nature and amount of manganese present in the soil and aerobic biological activity. Anaerobic conditions enhances Mn^{2+} availability and results toxicity. Under toxic conditions, leaf colour changes to dark blue.

C. Deficiency of nutrients

1. **Calcium and magnesium deficiency:** since the bases are lost during the formation itself, these soils face wide spread Ca and Mg deficiency.
2. **Molybdenum deficiency:** Since the soils are highly acidic Mo deficiency is most likely to occur at a pH < 5.6.
3. Nitrogen and phosphorus deficiencies are common and sometimes sulphur deficiency also is experienced in acid soils, especially when soil organic matter content is low. Phosphorus unavailability is a main problem under highly acidic condition due to P fixation by Fe and Al.

D. Low microbial activity: Bacterial population generally prefer a neutral to slightly alkaline environment restricting their activity under acidic conditions.Highly acidic soils inhibit the survival of useful bacteria. This may adversely affect N fixation by Azotobacter and Rhizobium and nitrification. Actinomycetes also prefer a neutral condition. Fungi will be the dominant group of microbes in acidic soils, often causing root diseases.

Legume nodulation and nitrogen fixation are frequently affected by one or more of the deficiencies or toxicities commonly associated with acid soils. Even though effective nodulation has occurred, acid soils limit nitrogen fixation by reducing growth of the host plant and thus in turn limiting the growth of nodule rhizobia. Low pH, Ca and Mo deficiencies, and Mn and Al toxicities have all been shown to affect nodulation. Mo deficient plants show symptoms of nitrogen deficiency while nodules are very small and numerous. Higher the soil Ca concentration, the lesser will be the effect of low pH and Mn and Al toxicities.

Amelioration of Soil Acidity

Management of soil acidity is needed for enhancing crop production and this is achieved mainly through the addition of liming materials. Lime has been recognized as an effective soil ameliorant as it reduces Al, Fe and Mn toxicity and increases base saturation, P and Mo availability of acid soils. Liming also increases atmospheric N fixation as well as N mineralization in acid soils through enhanced microbial activity. However, economic feasibility of liming needs to be worked out before making any recommendation.

Liming materials are generally added to acid soils for their reclamation by raising soil pH. Liming materials are the materials necessary for neutralisation of H^+ ions in soil solution. They are oxides, hydroxides, carbonates and silicates of Ca and Mg. The presence of these elements alone does not qualify it as a liming material, but the accompanying anion must be one that will reduce the H^+ activity in soil solution.

Reactions of lime in soil

When liming materials are added to soil, they react with CO_2 and H_2O to form bicarbonates which combine with H^+ ions and neutralize it. The reactions of different liming materials are presented below.

$$CaO + H_2O + 2CO_2 \rightarrow Ca(HCO_3)_2$$

$$Ca(OH)_2 + 2CO_2 \rightarrow Ca(HCO_3)_2$$

$$CaCO_3 + H_2O + CO_2 \rightarrow Ca(HCO_3)_2$$

$$Ca(HCO_3)_2 \rightarrow Ca^{2+} + 2HCO^-_3$$

$$H^+ + HCO^-_3 \rightarrow H_2O + CO_2$$

The liming materials on reaction with soil colloids, replace H^+ and Al^{3+} ions on the soil matrix by Ca^{++} ions. HCO_3^- ion neutralizes the H^+ ion present in soil solution and Ca gets adsorbed on the exchange complex.

$$\boxed{\text{Micelle}}\begin{matrix}H^+\\H^+\end{matrix} + CaO \rightarrow \boxed{\text{Micelle}}\,Ca^{++} + H_2O$$

Thus, addition of liming materials, neutralizes the H^+ ion in the soil solution and that present on the exchange sites. The reaction proceeds till H^+ ion concentration decreased. In acid soil, the H^+ ion concentration in soil solution depends on the hydrolysis of Al^{3+} or hydroxy aluminium or hydroxy ferric ions. The continued removal of H^+ion from the soil increases the soil pH.

Factors affecting liming reactions

1. **Moisture :** Greater the moisture content, more rapid is the rate of liming reactions.
2. **Temperature :** The preferred temperature range is between 20 and 25°C
3. **Exchangeable acidity :** Greater the exchangeable acidity, more rapid will be the rate of liming reactions

Common liming materials

1. **Calcite or lime stone ($CaCO_3$):** Obtained from shells or ground lime stone
2. **Dolomite : $CaCO_3 \cdot MgCO_3$:** Mined out by open pit method and are crushed to smaller size. Behaviour is similar to calcite and neutralizing value (NV) is 109.
3. **Quick lime or burnt lime (CaO):** Manufactured by roasting calcite by the removal of CO_2. It is the most effective liming material, with a NV of 179. For immediate results this should be used. While applying to soil, CaO should be thoroughly mixed with the soil to prevent carbonation.

4. **Marl :** It is mainly calcium carbonate or lime-rich mud or mudstone which contains variable amounts of clays and silt and present in lake bottoms. The dominant carbonate mineral in most marls is calcite, but other carbonate minerals such as aragonite, dolomite, and siderite may also be present.

5. **Chalk :** It is mainly $CaCO_3$, resulting from soft lime stone, deposited in oceans long time ago. It is a soft, white, porous sedimentary carbonate rock.

6. **Slags :** Three types of slags are available for use.

   a. **Blast furnace slag:** It is a by-product of pig iron industry. During the reduction of Fe, CO_2 is lost and Ca in the ore combines with molten silica to form a slag. It contains $CaSiO_3$ and $CaSiO_4$. NV is about 75-90.

   b. **Basic slag:** It is the by-product of basic Open-Hearth process for steel production. NV is about 60-70.

   c. **Electric furnace slag:** It is produced during the electric furnace reduction of phosphate rock for the manufacture of elemental P. NV is about 89.

7. **Miscellaneous materials:** Locally available materials like fly ash, biochar, rice husk ash, press mud, sludge from industrial water treatment plants etc. can also be used as liming materials. Liming materials of high-end technology like polymer coated lime or nano lime are being tested for their ability for acidity neutralization.

Though several materials are available for correcting soil acidity, the most important problem is their high cost which often prevents farmers from using them. Liming materials have to be selected based on their cost and efficiency.

Efficiency of liming materials

Liming materials differ widely in their efficiency. The efficiency of liming materials is evaluated on the basis of following important parameters.

1. Neutralizing value (NV) or calcium carbonate equivalent of liming materials
2. Purity of liming materials
3. Fineness of liming materials

Neutralizing value (NV) or calcium carbonate equivalent (CCE): It is the acid neutralizing capacity of agricultural liming materials expressed as a weight percentage of calcium carbonate

$$CCE = \frac{\text{Molecular weight of calcium carbonate}}{\text{Molecular weight of liming material}} \times 100$$

Table 2. Neutralizing value of common liming materials

Liming material	Chemical formula	Neutralizing value
Calcite	$CaCO_3$	100
Quick lime	CaO	179
Dolomite	$CaCO_3.MgCO_3$	109
Slaked lime	$Ca(OH)_2$	136
Calcium silicate	$CaSiO_3$	86
Magnesium carbonate	$MgSiO_3$	84

Purity: The effectiveness of a liming material is decided by its purity to a great extent. Presence of contaminants reduces the effectiveness of liming material.

Fineness of liming material: It indicates the rate of reaction of a liming material and is determined by the particle sizes of the material. The liming material that passes a 100-mesh sieve will react within a short period of time while only 60% of the liming materials that passes a 20-mesh sieve (but held on 100 mesh sieve) will react within that period (usually calculated time is one year). Material that does not pass the 20 mesh sieve is not expected to react within a year following application. Hence for practical use, limestone CCE equivalents need to be adjusted for the fineness of the material. Finer the material, it is more effective. For evaluating fineness, a "Fineness factor" is derived based on size of liming material. Neutralizing index is calculated by multiplying CCE with fineness factor.

$$\text{Neutralizing index} = \text{CCE} \times \text{Fineness factor}$$

Rating for fineness of liming materials

Material size	Efficiency rating (%)
Passes through 60 mesh sieve	100
Passes through 20 mesh but not 60 mesh sieve	60
Passes through 8 mesh but not 20 mesh sieve	20

An example for calculation of fineness factor is given below

Calculate the fineness factor for a liming material of which 50% passes through 60 mesh sieve, 30 % passes through 20 mesh but not a 60 mesh sieve and 20 % passes through an 8 mesh but not a 20 mesh sieve.

The fineness factor will be the sum of the products of the percentage of material in each of the three size fractions multiplied by appropriate fineness factor.

$$\begin{aligned}\text{Fineness factor} &= 50 \times 100/100 + 30 \times 60/100 + 20 \times 20/100 \\ &= 50 + 18 + 4 \\ &= 72\%\end{aligned}$$

Lime requirement

Liming materials should be applied as per the lime requirement. It may vary with the soil and crops. Certain crops like tea, citrus, pineapple, cassava etc. are acid tolerant while leguminous crops are highly susceptible to low pH. Similarly soils with high buffering capacity, (high CEC, high clay content, high in organic colloids) also require more liming material. The lime requirement is governed by four factors.

1. Required change in pH
2. Buffering capacity of the soil
3. Chemical composition of liming material and
4. Fineness of the liming material

Lime requirement is defined as the quantity of liming materials to be applied to the soil for increasing soil pH to a desired value, usually 6.5.

The lime requirement of a soil is determined by either SMP buffer method proposed by Shoemaker, Mc Lean and Pratt (1961) or on the basis of exchangeable Al content of soil (Kamprath, 1984). The amount of $CaCO_3$ required to neutralize a specific quantity of exchangeable Al is given by an equationas presented below.

$CaCO_3$ (tons ha^{-1}) = cmol ($1/3Al^{-3}$) kg^{-1} × Factor (Kamprath, 1984)

In exchangeable Al method, the factor for tolerant crops will be 1.0 while for medium tolerant plants it is 1.5 and for sensitive plants it is 2.0. For Ultisols and Oxisols, the factor will be in the range of 1.5 to 2.0 times of exchangeable Al in general.

In SMP method, 5 g soil is added to SMP buffer, equilibrated for 30 minutes and pH is measured. Based on the pH value the amount of lime to be applied can be calculated by looking into the standard table provided. All over India, based on the results of the network project on soil acidity, it is advised to apply one tenth of the lime required as per SMP buffer method for amelioration of acidity.

The desirable soil pH range for most of the field crops is 6.0-7.0. Lime requirement as per SMP buffer method of an acid soil in terms of pure calcium carbonate can be observed from the following table.

Table 3. Lime required for bringing soil to indicated pH as per SMP buffer method

pH of soil buffer suspension (field soil sample)	Lime required to bring soil to indicated pH (tons of $CaCO_3$ per acre)		
	pH 6.0	pH 6.4	pH 6.8
6.7	1.0	1.2	1.4
6.6	1.4	1.7	1.9
6.5	1.8	2.2	2.5
6.4	2.3	2.7	3.1
6.3	2.7	3.2	3.7
6.2	3.1	3.7	4.2
6.1	3.5	4.2	4.8
6.0	3.9	4.7	5.4

Application of lime

Application of small amounts of lime in every year or twice in a year is highly effective. Lime should be applied well in advance to the sowing and fertiliser application. Lime can be broadcasted to the soil and should be thoroughly mixed with soil. Economic and efficient liming materials may be selected for application.

Benefits of liming

1. Increases the N availability by providing a favourable environment for organic matter decomposition by microbial action
2. Increases P availability in acid soils
3. Make K utilization more efficient, by preventing excess uptake of K
4. Supply Ca and Mg
5. Make Mo available
6. Reduce the availability of Fe, Mn and Al and prevent their toxicity
7. Enhances root development and improves soil structure by enhancing the microbial activity and plant nutrient availability

Effects of over liming

Over liming also results adverse soil conditions for crop growth. Some of the important problems are

1. Deficiency of Fe, Mn and Zn
2. Reduced availability of P and K
3. Boron deficiency may occur
4. Root development may be hindered due to the dominance of OH^- ions
5. Incidence of Scab disease in root crops

Management of Acid Soils

Acid soil management is being carried out with the objective to improve crop yield and maintain soil in a healthy condition. This can be accomplished either through the addition of amendments or by manipulation of management practices and cultivation of acid tolerant crops/varieties. However, the most common practice is the use of amendments where liming materials are added to soil in advance to sowing. Sufficient time has been provided for the acidity neutralization reactions to take place. The increase in pH will favour the satisfactory growth of seedlings since it improves base status, nutrient availability and reduces the activities of Fe, Mn and Al. While choosing the liming material, efficiency and cost are the two important factors to be considered. A wide range of liming materials are available ranging from naturally occurring materials like calcite /dolomite to industrial byproducts. Lot of research is going on new formulations like pelletized lime, polymerized lime, biochar blended lime etc. for their efficacy in correction of soil acidity, but the results are yet to come. Most of the acid soils are low in fertility and productivity. Hence judicious use of efficient liming materials, manures and chemical fertilizers alone could provide a good crop yield from acid soils without losing soil health.

Acid tolerant plant species or varieties can be selected for cultivation. Plants like cassava, pineapple, banana etc. can tolerate soil acidity to a certain extent.

Application of organic manures regulates the soil pH by affecting the buffering capacity of soil. Regular application of organic manures like FYM, vermicompost, green manures etc. will improve the physical, chemical and biological properties of soil. Application of ash also reduces soil acidity. Use of rice husk ash for correction of acidity in paddy fields is being promoted by farmers in Kuttanad, Kerala.

Among the acid soils, acid sulphate soils and laterite and associated soils need special reference and are detailed below.

2. ACID SULPHATE SOILS (CAT CLAYS)

Acid sulphate soils are unique soils containing iron sulphides, that have formed naturally in water-logged conditions such as estuaries, wetlands and shallow groundwater in deep sands. Soils containing mineral sulfides (mainly pyrite) that become very acid on drying have been recognized years back and its first mention was done by Dutch farmers who named this infertile soil as *Kattakali* meaning Cat Clays. Chenery (1954) introduced the term "Acid sulphate soils" that have been drained, have adsorbed sulphate and pale yellow colour of Jarosite and usually have low pH (< 4.0) when in water. If the soils are drained or exposed to air by a lowering of the water table, the sulphides react with oxygen and form sulphuric acid. Undrained soils with high content of pyrite which have the potential to become more acidic on drainage also occur in situations where some forms of reclamation have been attempted. These are termed as "potential acid sulphate soils".

Acid sulphate soils are generally formed by natural processes during the Holocene geological age (the last 10,000 years). They were originally deposited in marine, estuarine or riverine sediments and occur predominantly in low-lying coastal floodplains, rivers and creeks, deltas, coastal flats, back water swamps and mangrove forests and in coastal areas subjected to sea water inundation. Metal sulfides can also be found in many rock types, generally at low concentrations. Acid sulphate soils are found worldwide except in Australia. They occur in all climatic zones ranging from cold-temperate to humid tropics. They are found most widely in tropical deltas. In India, acid sulphate soils are found in Kerala, Orissa, Andhra Pradesh, Tamil Nadu and West Bengal. These soils belong to soil subgroups like Sulphaquepts / sulphihemists / sulphohemists / sulphaquents.

Most of the acid sulphate soils are poorly drained because of the unripe nature of soil, or nearness to sea / shallow water table. Sulphidic materials or metal sulphides (FeS_2 and FeS) are present, which on oxidation produces sulphuric acid and liberate colloidal Al and Fe, decreasing the soil pH to below 3.5. Drainage of peaty acid sulfate soil material also results substantial production of the greenhouse gases CO_2 and N_2O. Infrastructure constructed on acid sulphate soils often gets corroded.

The oxidation of metal sulfides is a function of natural weathering processes which is generally slow. But the soils get rapidly oxidized when they are exposed to air by human activities / mining activities. Soil horizons that contain sulfides called as 'sulfidic materials' can be environmentally damaging if exposed to air by disturbance. Exposure results in the oxidation of pyrite. This process transforms sulfidic material to sulfuric material and the pH decreases to 4.0 or less.

Identification of Acid Sulphate Soils

Acid sulphate soils are identified based on the presence of sulphidic / sulphuric materials and pH of soil : water suspension. The general criteria for identifying an acid sulphate soils are the following

1. Presence of sulphuric horizon which has a pH (1:1) of <3.5 and evidences of presence of sulphide content (Yellow colour)
2. The pH of soil water suspension (1:2) should be < 4.0.
3. Presence of yellow mottles of jarosite ($KFe_3(SO_4)_2(OH)_6$) in oxidized (Eh $>$ 400 mv) condition. But under submergence pH might be slightly higher and jarosite get reduced.

These soils are generally strongly acidic containing toxic quantities of Al, Fe, S, Mn and soluble salts (unless leached).

Classification of Acid Sulphate Soils

They are classified into two

1. True acid sulphate soil and
2. Potential acid sulphate soil

True acid sulphate soil possesses a highly acidic soil layer containing sulfuric materials produced by the oxidation of sulfidic materials present. This layer is known as sulphuric horizon. Sulfuric material is composed either of mineral or organic soil material (15 cm or more thick) that has a pH <3.5 and can usually be identified by the presence of bright yellow jarosite mottles or streaks.

Potential acid sulphate soil is that, which has not been exposed to air or undergone oxidation but has the potential to develop actual acid sulphate condition when exposed to air or drained. It is composed mostly of accumulations of iron sulfide minerals. Iron sulfides can react rapidly when they are disturbed (exposed to oxygen). Pyrite will tend to occur as more discrete crystals in soil and organic matter matrices and will react more slowly when disturbed.

Apart from the true and potential acid sulphate soils, another category viz., para or pseudo acid sulphate soils are also found. These soils have a pH >4.0 and contains basic iron sulphate. Para or pseudo acid sulfate soils are soils in which the acid has been leached out or neutralized to the extent that microbiological activation and root development are no longer hampered and which still show jarosite mottles, high soluble sulfate and high percentage of Al saturation of the clay. Such a soil contains one or more horizons with the characteristic yellow mottling (basic iron sulphates) commonly associated with acid sulphate conditions, but does not have a pH below 4.0 and does not contain free acids or more than about 60 per cent exchangeable Al.

Origin and Genesis of Acid Sulphate Soils

The formation of acid sulphate soils takes place only under certain conditions. The pre-requisites for the formation of acid sulphate soils are

1. **Source for sulphur :** Sulphur is the major element that undergoes reduction and oxidation vice versa along with iron which forms the characteristic feature of the acid sulphate soils. The sea water which is rich in sulphates of various bases (Na / Ca / Mg) serves as the source of sulphur. Mangroves the natural vegetation, are also rich in sulphur content. The decomposition of mangrove leaves and the sea water provides sufficient quantity of sulphur.
2. **Presence of excess quantities of iron :** The presence of pyrite in reduced situation and of jarosite under oxidized condition is essential for the development of acid sulphate soils. The soils inherently rich in iron are present in coastal belts of tropics.
3. **Supply of organic matter :** Organic matter is the source of energy for the reaction involving the genesis of acid sulphate soils. The organic residues in varying stages of decomposition derived either from the mangrove vegetation or vegetation of backwater swamps provide ideal condition for the development of acid sulphate soils. The dense vegetation speed up the pyrite formation by the renewal of dissolved sulphates and removal of soluble byproducts through tidal flushing.
4. **Development of acid sulphate soil following drainage :** Pyrite is stable only under anaerobic conditions. When drainage is given, it gets oxidized, generating sulphuric acid.
5. **Anaerobic or flooded condition :** The nearness to sea / backwater swamps / deltas provided an anaerobic situation due to the presence of shallow water table or flooded situations. In a flooded saline or brackish water tract, with ample supply of sulphur, iron and organic matter, there is every chance for the formation of acid sulphate soils.

Genesis of acid sulphate soils

Genesis of acid sulphate soils includes the following processes.

1. Cumulative and reductive geochemical phase
2. Oxidative phase followed by
3. Neutralization phase

1. **Cumulative and reductive geochemical phase:** This is the reductive phase under which pyrite formation and its accumulation occurs in soil. This phase includes pyrite formation which involves bacterial reduction of sulphates to sulphides, partial oxidation of sulphides to elemental sulphur and disulphides, and interaction between ferrous or ferric iron with sulphides and elemental sulphur. The accumulation of pyrite is a combined effect of bacterial and chemical reactions. Bacteria breaks down the organic matter under anaerobic conditions, reduces sulphates from sea water to sulphides and Fe(III) oxides to Fe(II) oxides. The rich vegetation that gets submerged in sea water on decomposition and enrichment with sulphates from sea water leads to the formation of sulphides.

$$SO_4^- + 10H^+ \rightarrow H_2S + 4H_2O$$

$$2H_2S + O_2 \rightarrow 2S + 2H_2O$$

$$Fe^{++} + S \rightarrow FeS$$

$$FeS + S \rightarrow FeS_2 \text{ (Iron disulphide)}$$

Thus, the accumulation of pyrite is brought about by the combined effect of unique condition that exist in tropical coastal areas. The sulfur in pyrite is derived from the sea water, which is biologically reduced to sulfide in the anaerobic mud. An energy source is necessary for bacterial sulfate reduction, and organic matter serves this purpose. It is readily available as a result of abundant plant growth in these coastal areas. Large quantity of ferrous iron (Fe^{2+}) is also available, usually by the reduction of insoluble ferric compounds that results from the weathering of clay.

The combination of sulfates from sea water, organic matter from plant growth, the anaerobic condition due to submergence and the presence of Fe^{2+} result in the formation and accumulation of pyrite in tropical coastal wetlands. FeS formed get oxidized to (with Fe (III) and O_2 as oxidants) to disulfide (S_2^{2-}). Pons *et al.* (1982) proposed that the solid-solid reaction of FeS and S to form FeS_2 is a slow process, which takes months or years to produce measurable quantities of pyrite; but the direct precipitation of Fe^{2+} and S_2^{2-} to form FeS_2 yields pyrite within days under favorable conditions (Goldhaber and Kaplan, 1974).

The overall reaction is presented below which includes reduction of all sulfate to sulfides.

$$Fe_2O_3 + 4SO_4^{2-} + 8CH_2O + 1/2O_2 \rightarrow 2FeS_2 + 8HCO_3^- + 4H_2O$$

2. **Oxidation of Pyrite:** The fine-grained pyrite typical of tidal sediments gets readily oxidized upon exposure to air, giving Fe(II) sulfate and sulfuric acid: The rate of oxidation is enhanced by *Thiobacillus ferrooxidans* which is active under pH below 3.0.

$$2FeS_2 + 7O_2 + 2H_2O \rightarrow 2Fe^{2+} + 4SO_4^{2-} + 4H^+$$

Complete oxidation and hydrolysis of iron to Fe(III) oxide yields 2 moles of sulfuric acid per mole of pyrite.

$$FeS_2 + 15/4\ O_2 + 7/2H_2O \rightarrow Fe(OH)_3 + 2SO_4^{2-} + 4H^+$$

(van Breemen, 1982).

Pyrite is oxidized more rapidly by dissolved Fe(III) than by oxygen, according to the following reaction

$$FeS_2 + 14Fe^{3+} + 8H_2O \rightarrow 15Fe^{2+} + 16H^+ + 2SO_4^{2-}$$

The reaction of pyrite with oxygen is slow, but it gets rapidly oxidized by Fe^{3+} in solution and Fe^{3+} get reduced to Fe^{2+}. At the same time Fe^{3+} is regenerated from Fe^{2+} by the bacteria *Thiobacillus ferrooxidans*. This catalytic oxidation takes place only below pH 4.0 because Fe^{3+} is soluble at that pH only. Another group of bacteria (*T. thioxidans*) is involved in oxidation at higher pH.

Fe^{3+} ultimately crystallizes as reddish brown oxide "goethite" in mottles / nodules in soil. Some iron lost in water and may get deposited at other places and often block the drains.

3. **Neutralization phase:** Sediment containing pyrite becomes a potential acid sulphate soil only when the potential acidity exceeds the neutralization capacity of the soil. One part of pyrite S is neutralized by three parts of $CaCO_3$.

   The sulphuric acid formed during oxidation reacts with bases in sediments and if the base content of the sediment is small extensive acidification of soils takes place. Several oxidation products are formed under this stage. The most prominent compounds are jarosite, oxides of iron and gypsum. They further undergo various reactions in the soil. The followings are some of the important oxidation products.

Oxidation products

1. Jarosite: Jarosite is formed only in extremely acidic (pH 2.0 to 4.0) and oxidized (Eh >400 mv) environments. The pale yellow (2.5-5Y 8/3-8/6) jarosite ($KFe_3(SO_4)_2(OH)_6$) is conspicuous in most acid sulfate soils. It commonly occurs as earthy fillings of voids or as mottles in the soil matrix. In acid sulfate soils, the jarosite is metastable and will eventually be hydrolyzed to goethite. The pale yellow mottles are so characteristic that they are used, together with pH, as a diagnostic criterion for classifying acid sulfate soils. However, jarosite is lacking in some acid sulfate soils, particularly those high in organic matter.

2. **Iron Oxides:** Most of the iron from oxidized pyrite ends up as Fe (III) oxides. Fine-grained goethite may form either directly, and quickly, upon oxidation of dissolved Fe(II) sulfate released during pyrite oxidation, or more slowly, by hydrolysis of jarosite. The reaction follows.

   $Fe^{2+} + SO_4^{2-} + 1/4O_2 + 3/2H_2O \rightarrow FeOOH + 2H^+ + SO_4^{2-}$

   $KFe_3(SO_4)_2(OH)_6 \rightarrow 3FeOOH + 2SO_4^{2+} + K^+ + 3H^+$

In the better drained, deeply developed acid sulfate soils, part of the Fe (III) oxides in the B horizon may occur as hematite, giving conspicuous red mottles.

3. **Gypsum:** Calcium carbonate is the prime agent for neutralizing the acidity. Appreciable amount of $CaCO_3$ if present in soil, the precipitation reaction could occur as follows.

$$2CaCO_3 + KFe_3(SO_4)_2(OH)_6 + 5H_2O + H^+ \rightarrow 2CaSO_4 \cdot 2H_2O + 3Fe(OH)_3 + K^+ + 2CO_2$$

The formation of gypsum in acid sulfate soil is an indication of the soil being relatively suitable for agriculture.

Fate of acidity

The acidity produced is neutralized by $CaCO_3$ present in the soil or by added limestone, finally producing gypsum.

$$CaCO_3 + 2H^+ + 2SO_4^+ + H_2O \rightarrow CaSO_4 \cdot 2H_2O + CO_2$$

Some amount of acidity can be neutralized by the bases present in the soil and a part by flooding and leaching. During flooding acidity is alleviated due to the consumption of H^+ as shown below. It also alleviates Al toxicity. But Fe toxicity still persists since iron is soluble at pH above 5.0.

$$SO_4 + 2H^+ + \underset{\text{(Org. matter)}}{2CH_2O} \rightarrow H_2S + 2H_2O + 2CO_3$$

$$Fe(OH)_3 + 2H^+ + 1/4CH_2O \rightarrow Fe^{2+} + 11/4H_2O + 1/4CO_2$$

At pH values below 3.5, Al toxicity is the principal hazard, though Fe and H are present in toxic quantities. H_2S toxicity is also prevalent as a result of sulphate reduction, but the presence of Fe^{2+}, reduces its intensity by forming iron sulphides.

Constraints for Crop Production

1. Soil acidity: Presence of sulfuric horizon or sulphidic materials results a very low pH imposing toxicity due to direct effect of H^+ ions. These soils are generally formed as a result of drainage of parent

materials that are rich in pyrite. Pyrite accumulation takes place in waterlogged soils that are rich in organic matter and are flushed by sea water containing dissolved sulphates. When the pyrite is exposed, it gets oxidized to sulphuric acid and pH drops to below 4.

2. Salinity : Another problem is salinity since most of these soils are located in back water areas or very near to saline water bodies. Hence the cultivation in acid sulphate soils has to be initiated only after addressing both acidity and salinity problems. Hence rice varieties that can tolerate both acidity as well as salinity are to be cultivated. In certain areas where there is direct access to sea water by tidal action, the extent of salinity is very high and saline tolerant varieties alone are cultivated and that too in seasons when the salinity will be minimum. Eg. rice varieties cultivated in *Pokkali* soils of Kerala, India.

3. Chemical toxicities: Extreme acidity and associated iron, aluminium and sulphur toxicities are quite common in acid sulphate soils. Certain extent of acidity can be eliminated by flooding, but it results Fe, Al and H_2S toxicity. In old acid sulphate soils where pyrite oxidation is complete, flooding can remove a good amount of acidity.

4. Low nutrient content: These soils are mainly deficient in phosphates and bases like Ca, Mg, K, Zn, Cu, though these areas are usually inundated by sea water. The presence of excess quantities of Al, Fe and H of course had antagonistic effects on other bases. The extent of deficiency may vary with parent material or other factors of soil formation.

5. Physical limitations: Root development is restricted. The extreme toxicities present even prevent root initiation, especially by Al and the developed roots may get decayed or get deposited by iron ochre, causing only limited root area utilized for nutrient absorption. Poor root growth badly affects the vegetative and reproductive growth of paddy. Water reserves in subsoil are not available. Arrests soil ripening and maintain soil in soft fluffy condition with shallow depth.

Risks Associated with Acid Sulphate Soils

During excavation of acid sulfate soils, care should be taken not to expose the sulfidic materials to air. If somehow it gets exposed, appropriate disposal of the excavated material should be practiced. During dewatering or lowering of the water level, there may be chances for the exposure and acidity may get intensified. Generally, any proposal to disturb soils or interfere with the water table where potential acid sulfate soil materials or actual acid sulfate soil materials exist should bear this in mind.

Such exposures usually occur when acid sulphate paddy soils are converted to aquaculture ponds. While digging, if the sulfuric horizon gets exposed, large amount of acidity is released to the water. The aquatic organisms or plants will not be able to thrive under these situations. Such degraded acid sulphate soils need neutralization of acidity and require several years to regain the equilibrium.

Impacts of Acid Sulfate Soils on Environment

Acid sulfate soil materials if present in any area would definitely affect land use, development and amenity of the surrounding environment. The major impacts are on the following :

1. **Engineering and landscaping works:** Sulfuric acid generation can result in the corrosion of concrete, steel and some aluminium alloys used in buildings, drainage systems and roads. Hence choose suitable structures that can resist / withstand acidity for such areas. If acid sulfate soil materials are used as site fill material or in embankments, will affect plant growth adversely and block pipe drainage systems due to the formation of iron oxides.
2. **Agricultural practices:** Both acidity and increased liberation of soluble metals results in direct plant toxicity and decreases the availability of some nutrients resulting a decline in farm productivity. The animal productivity also will be affected due to decrease in pasture quality and an increased uptake of aluminium and iron by grazing animals.

3. **Fish and aquatic life:** Fish / crustaceans mortality due to the discharge of acidic waters to estuaries / coastal or riverine environments is common in acid sulphate tracts. The survival of many of the aquatic plants is also very difficult due to direct acid exposure and toxicity by iron, aluminium and heavy metals.
4. **Local amenity:** The presence of acid sulfate soil materials produces an offensive odour, predominantly due to 'rotten egg gas' (H_2S). Submerged saline environment carrying organic matter under different stages of decomposition also emits bad odour.

Reclamation of Acid Sulphate Soils

1. **Submergence :** Keep the soils in flooded condition as long as possible: Adopt flooded rice cultivation with minimum disturbance on soil so that the exposure of soil to air and consequent oxidation and generation of acidity can be avoided.
2. **Leaching and drainage :** The high concentration of soluble salts and aluminium can be reduced by leaching. Provision of shallow surface or subsurface drainage will facilitate the leaching of acid produced due to oxidation of pyrite.
3. **Application of amendments :** Liming materials to neutralize soil acidity temporarily for a short period are applied to the soil at the time of land preparation and the acidity released to soil are washed away. This will facilitate rice seeds germination and survival of seedlings in acid sulphate soil.
4. **Crop nutrition :** Judicious fertilizer recommendation based on the soil test data will be able to correct nutrient deficiencies and promote growth.
5. **Practice rice - shrimp cropping sequence :** Rice is grown in the wet season when the land is flooded by freshwater. After the harvest of rice, brackish tidewater is lets into the fields in the dry season to raise a valuable shrimp crop. This system keeps the sulphidic material flooded to stop generation of acid and raises the pH by reduction.

6. **Application FYM / indigenous amendments :** Application of FYM or other composted manures helps the rice seedlings to overcome the initial toxicities and yield satisfactorily. Application of biochar / rice husk ash well in advance to sowing / planting increased the soil pH and reduced the toxicities.
7. **Tolerant or resistant varieties :** Rice is the major crop cultivated in acid sulphate soils. Several rice varieties had been identified / developed by agricultural research institutions to suit adverse soil conditions. Cultivation of site specific suitable varieties can reduce the problems.

3. LATERITE AND ASSOCIATED SOILS

Laterite and associated soils are distributed in the tropics and subtropics of Africa, India, South America and South East Asia. In India these soils are found in southern, south eastern and north eastern states (AER 8, 18 and 19). Well developed laterites are found in hill tops and plateaus of Kerala, Tamil Nadu, Orissa and sparingly in Andhra Pradesh and Karnataka. Lateritic soils are widely distributed in all the above states.

3.1 Laterite Soils

In India, laterite soils are found in Kerala, Deccan hills, Karnataka, Tamil Nadu, Orissa and Assam covering an area of 130066 sq. km and are well developed on the summits of Deccan hills, Karnataka, Kerala, Eastern Ghats, west Maharashtra and central parts of Orissa and Assam (Varghese and Byju, 1993). The term laterite (Latin word *later* = brick) was first coined by Dr. Francis Hamilton Buchanan in 1807 for ferruginous, vesicular and unstratified soft materials occurring within the soil which hardens irreversibly on exposure to air, found in the North Malabar, Kerala. As per soil taxonomy laterite is replaced by the term "Plinthite" (Greek word *plinthos* meaning brick). It is highly weathered material enriched with secondary form of Fe and Al, devoid of bases and primary minerals. It is full of pores and cavities and contains large quantities of iron in the form of red and yellow ochre. In the mass, excluded from air, it is soft and can be cut to make brick stone by any sharp instrument, which later hardens on exposure to air.

Laterite soil is the soil in which a laterite subsurface horizon is present. Laterite soils are formed by intense leaching of bases and desilication followed by residual accumulation of iron and aluminium oxides. The main components of these soils are silica, alumina and iron oxides with small amounts of titanium and manganese oxides. Laterite soils are having a silica:alumina ratio of <1.33 and soils with values above this is designated as lateritic soils (Martin and Doyne, 1927). Kellog (1949) used the term Latosols to designate lateritic soils and the term laterite is restricted to ferruginous material which hardens on exposure to air. While Robinson (1949) used the term "ferrallitization" for the pedogenesis and "ferrallitisols" for the resulting pedon. However, the hardening is an essential criterion for a soil being designated as laterite. Laterite soils are formed under humid tropical climate experiencing distinct alternate wet and dry spells. Soil Survey Staff (1975) introduced the soil order "Oxisols" to cover the laterite soil where the diagnostic characteristic is an Oxic subsurface (B) horizon. However, most of the laterite soils classified under Ultisols since the conditions do not match with the specifications mentioned in Keys to Soil Taxonomy.

Two forms of laterites are found in India.

1. **High level laterites (Primary laterites):** Formed by *in situ* weathering and found to cap the summits of hills and plateaus in the central and western parts of India. They never occur below 2000 ft.
2. **Low level laterites (Secondary / detrital laterites) :** Formed by the transportation of laterites present in higher topographic positions to downwards by streams and again cemented into compact mass, including grains of quartz and other minerals. In most of the cases they are secondary in origin and are derived from high level laterites and recombined after deposition in the valleys. Thus, they are detrital in origin and are found in the large tracts of Deccan peninsula and Kerala.

Genesis of Laterite Soils

Laterite soils are formed in humid tropics as a result of the process called laterisation. Particular humid tropical climatic conditions with heavy precipitation followed by a continuous dry spell of greater than 90 days is essential for the initiation of laterisation. The factors affecting laterisation are given below.

1. **Climate :** Two most important climatic parameters are precipitation and temperature. Laterisation will take place only in humid tropical areas with distinct alternate dry and wet spells. A high precipitation, of more than 1500 mm within the monsoon season followed by a dry spell for more than six months is ideal for the formation of typical laterite soil. High temperature and precipitation of humid tropics favours the laterisation process. They are generally found in tropics characterized by a soil temperature of 24 to 29°C. Variation in the above specifications may result the formation of ferrallitic / associated soils.
2. **Topography :** Laterite soils are found in flat surfaces, gentle slops and in undulating and rolling terrain wherever good natural drainage is assured.
3. **Parent material :** Laterite can be formed over a wide range of parent rocks ranging from basic basalt to acidic granites. Ferruginous laterites may develop on a variety of materials provided there is source of iron either in the parent rock or in adjacent higher lying areas from which water may introduce ferruginous material. The laterite layers may be in material unrelated to the bed rock. The high precipitation falling on parent basic rocks dissolves the bases in water removes it from the weathered material allowing only the materials resistant to alkaline hydrolysis to remain there. According to Buringh (1970), acid parent material results in the formation of kaolinite while the basic materials result in gibbsite and the iron content of parent materials largely decides the sesquioxide content.

4. **Ground water level :** Ground water level near to the surface will facilitate the hydration of Fe and Al resulting the formation of their hydrous oxides.
5. **Biosphere :** Role of vegetation, termites, earthworms and even human activity had their on specific roles in the formation of laterite soil.
6. **Time :** The role of time on laterite soil development is as important as in any other case. Many of the existing laterites are clearly the relics of geologic antiquity. The formation of a laterite soil from a solid unweathered rock is a slow process involving formation of unconsolidated weathered material which later get enriched with iron and aluminium followed by a drying crystallization phase leading to hardening or induration.

Laterisation takes place very slowly under ambient conditions. Based on the variation in climatic parameters like temperature, duration and intensity of rain fall, topography, nature of parent material etc. decides the rate and extent of laterisation. The steps of the laterisation processes are the following.

1. **Dissolution and elimination of silica, alkali and alkaline earth metals :** Under humid tropics, there is scope for intense weathering. The parent rocks are subjected to alkaline hydrolysis which leads to the formation of alkalis. Under alkaline hydrolysis, silica gets washed down and thus eliminated from the upper horizon. Along with silica, the soluble bases are also removed. Due to this process the silica:alumina (SiO_2:Al_2O_3) and silica:sesquoxide (SiO_2:R_2O_3) ratios decreases to < 2.0. Whatever organic matter is there, will also get leached out during the process.
2. **Accumulation of hydrated oxides of iron and aluminium :** Due to progressive leaching, bases and silica are taken out of the weathering rocks and the resistant metals like iron and aluminium get accumulated. Hydration leads to accumulation of hydrated oxides of iron and aluminium in some parts of B horizon.

3. **Induration or hardening of laterite :** During the dry spell, iron and aluminium hydroxides get dehydrated and subjected to crystallization which result the formation of an indurated or hardened material called as laterite or plinthite. The hardening is a complex phenomenon which includes two processes ie., dehydration and crystallization, apart from gross or local enrichment. The iron may combine with or get mixed with clay, and aluminium may partly combine with remaining silica. Dehydration is the prime factor that decides the development of hard laterite crusts. Hydrous amorphous oxides of iron lose water and crystallize as goethite and aluminium hydroxides to gibbsite, which is a stable mineral. With aging, goethite content decreases and hematite increases. Hardness also increases with aging. The ultimate stage of these processes leads to the formation of "latreite" which is no longer a soil, but indurated iron crust material otherwise known as "plinthite".

Another process that always compared with laterization is podsolization which occurs in humid temperate conditions. Under low humidity and temperate conditions, after autumn season, there will be a natural accumulation of decomposing leaves leading to high organic matter content in the soil. Due to microbial activity the organic matter get decomposed, yielding different organic acids, which initiate an acid leaching. During this process, the iron and aluminium get washed down leaving the silica at the soil surface, imparting a bleached appearance to the surface soil layers. At the lower layers the transported colloidal sesquioxides get cemented with sand grains forming a hard layer known as "orstein" above which usually the rust brown horizon rich in organic matter (humus) can be seen.

Characteristics of Laterite Soils

Morphological characteristics

The laterites identified in field consist mainly of a superficial cuirass of oxides of iron and aluminium. The massive form of laterite may either show vesicular or conglomerate structure. The vesicular structure is associated with the activities of termites. The termite channel allows air to get into the subsoil and so provide conditions necessary for the deposition of ferric hydroxide from ferrous solutions. The surface crust of laterite is indurated due to the changes in degree of hydration of oxides of iron and aluminium present.

The main visible feature of laterite is the accumulation, form, colour and consistency of iron hydroxides and oxides which impart yellow, pink, brown and red colours (5YR or redder) to the ground matrix and earthy clay. The structures of iron oxides are usually referred as concretionary, pisolitic, vesicular etc. The shape of these structures is related to topography.

These soils have high clay content especially in B horizon due to *in situ* alteration of weatherable minerals and illuviation. Dominant clay mineral is kaolinite. Goethite and gibbsite are also present.

In plateaus where top soil is eroded, the hard laterite is exposed and in gentle slopes, there is a soil cover of 15-30 cm depth or even up to 2 m. A crumbly layer of laterites with irregular blocks follows the surface layer. Beneath this, the laterite layer that can be used as building stone is present. A very soft laterite material is present below the above layer.

Laterite rock outcrops are commonly found in the plateaus, cliffs and isolated hills where portions of freely exposed bedrock protrude above the soil level due to natural reasons, known as duricrusts or ferricretes. Each of which is known to harbour highly specialized vegetation rich in habitat specific and endemic plants. The micro-environment at the rock surface ranges from very hot and arid in dry season to waterlogged in wet season. Soil formation on the outcrops is extremely slow. Soil depth varies from a few centimeters on flat areas to about a meter in deep cracks and depressions. It is sandy to sandy loam in texture, highly acidic and poor in phosphates. In the dry period, the temperature of exposed rock surface is very high and humidity is very low. In the monsoon, conditions go to other extreme when the rains are continuous and heavy, leading to formation of ephemeral wetlands. Scarcity of soil and extremes of microclimate are unfavourable for the growth of perennial vegetation. Hence, vegetation dominated by annuals thrives on outcrops only in moist conditions

Laterite has the peculiar property of being soft when newly quarried, but being hard and compact on exposure to the air; also, loose fragments and pebbles of rock tend to re-cement themselves into solid masses as compact as the original rock. On account of this property, it is usually cut in the form of bricks for building purposes. Laterite brick is generally red in color. It is porous and shows vermicular structure. In the areas of extensive laterite formations, its mining has emerged as a major economic activity of the local people All the laterite are not mineable, only those have been formed under the alternate spell of dry period for more than 90 days and with heavy rain for 6 months, with the other factors favourable for laterisation alone will provide good quality laterites. Others will be subjected to crumbling on exposure and could not get compact materials.

Chemical characteristics

Mature laterites are made up primarily of iron, alumina, silica, titanium and water. Sesquioxides are the major constituents followed by kaolinite. It is poor in organic matter, alkali and alkaline earth metals. Soil reaction is acidic with a pH range of 4.5 to 6.0. The CEC is also very low ranging from 2.5 to 7.0 cmol kg^{-1}. It contains the following minerals.

1. Iron oxides –Hematite, hydrated goethite, lepidocrocite and magnetite
2. Aluminium oxides- Gibbsite, boehmite and diaspora
3. Silica- In a typical laterite silica content will be very low
4. Manganese oxides- Lithiophorite and birnessite
5. Titanium oxides – present in very small quantities
6. P_2O_5- Because of immobility it accumulates in the surface soil

Fully developed laterites are poor in Ca and Mg. Na and K contents are also very low since these are easily leached out as laterisation proceeds. Minute quantities of Cr and Va are also detected in laterites.

Constraints

Laterite soils possess physical, chemical and biological constraints for crop production.

1. **Physical constraints:** Limiting factors are susceptibility to erosion, surface crusting, low water holding capacity, drought stress, excess gravel content etc.

    a. Poor water holding capacity: Water holding capacity is very low due to the dominance of kaolinite clay which is not an expanding type clay. The dominance of kaolinitic clay make soil light textured with hardened surface and low vegetation, making it more erosion prone. Reduced soil volume due to concretions and occurrence of plinthite and petro plinthite, and removal of surface soil by erosion and exposure of laterite bed results root zone limitations also.

    b. Drought stress: The very low water holding capacity of the soil's results water stress for plant roots.

    c. Gravel content: The gravel content of laterite soils varies widely. The gravelliness factor depending on the soil texture increases the bulk density which adversely affects the foraging capacity of plant roots for nutrients. Laterite soils developed under well distributed rainfall alone are without gravels and as the dry spell increases, there is an increase in gravel content also.

2. **Chemical constraints:** Laterite soils exhibit wide variation in relation to their origin, but this has not reflected much on soil properties. Most of the laterite soils have common problems. They are presented below.

    a. Deficiency of organic matter and nitrogen: Since these soils are highly weathered and leached, and mainly found in humid tropics where rate of organic matter decomposition is very high, naturally the organic matter content is very low. Nitrogen being mainly organic in origin wherever the organic matter content is low, nitrogen status will also be poor. Only in forest ecosystems the laterite soils show higher organic matter status.

b. Deficiency of phosphorus due to fixation: Due to the abundance of iron and aluminium, the P ions readily reacts with iron and aluminium forming their insoluble phosphatic compounds, which is commonly known as P fixation.

c. Deficiency of potassium, calcium, magnesium and sulphur: The alkaline hydrolysis had taken away all most all the bases and soluble ions from the soil, making it very infertile in nature.

d) Deficiency of micronutrients: Micronutrients like Zn and B are deficient. Zn is available at pH around 6.0 and many of the laterite soils are having pH below 5.5. The loss of boron from the soils by leaching and crop uptake warrants the application of boron containing fertilisers in laterite soils.

e) Very low cation exchange capacity: Due to the dominance of kaiolinitic clay and the low status of organic matter make these soils very poor in CEC. Since most of bases are lost during the soil development, the laterite soils record a base saturation less than 35 per cent.

f) Acidic pH and toxicity of iron, aluminium and manganese: The pH is always acidic ranging from moderately acidic to extremely acidic. Surface as well as subsoil acidity is prominent in laterite soils. In subsoils the pH may be 4.5 or less. Under low pH, elements like Fe, Al and Mn get solubilized and that present in insoluble forms became available resulting their toxicity. Fe toxicity is widely observed in paddy soils and the characteristic symptom is bronzing. The Al toxicity is also widely prevalent in paddy soils expressed as root decay and poor root growth, inhibits the root initiation and cause stunting of primary roots.

3. **Biological constraints:** The organic matter content of laterite soil is low. The high temperature and destruction of vegetative cover has accelerated the oxidative process due to exposure leading to depletion of the organic carbon stock. Hence the microbial activity is also less. Soil microflora utilizes organic matter as their food source and in soils with low organic matter the biological activity will be less. Since the soils are highly acidic, the fungi is be the dominant species.

Management Options

1. **Selection of suitable species and varieties:** Several crops are tolerant to soil acidity, especially the crops like tea, pineapple, coffee, cassava, cashew, banana, vegetables, coconut etc. These crops perform very well in laterite soils. For crops like rice, acid tolerant varieties have been developed.
2. **Liming:** The acidic pH and deficiency of calcium and magnesium in laterite soils are considered as the major problems for crop production. Liming with suitable amendments will improve the chemical characteristics of laterite soil and improve productivity. Practice liming according to the lime requirement of soil. Use of dolomite will address the Mg deficiency also.
3. **Fertilization:** Judicious use of manures and fertilizers as per soil test data are needed to tie over the low fertility status of soil. Special care has to be given for phosphorus management in laterite soils. Practice of integrated nutrient management system which could sustain the productivity of these soils is a good option. Management practices like foliar application of deficient nutrients, fertigation, precision farming etc. can be followed to meet the nutrient requirement of crops.
4. **Adoption of suitable soil conservation methods against erosion:** Erosion being the most important physical constraint, suitable agronomic or engineering soil conservation methods are to be followed.
5. **Breaking of hardpans:** The presence of hardpans will restrict root and water entry and make the plant growth very poor. Breaking of hardpans or deep ploughing will help to overcome this problem.

3.2 Lateritic Soils

Lateritic soils are the ones that do not necessarily have laterite as a subsurface horizon but reflects its occurrence in their morphological features. These soils will have silica:alumina ratio < 1.33 and silica:sesquoxide ratio > 2.0. Lateritic soils are also developed under conditions comparable to that of laterite soils, but such distinct wet and dry spells are not required for the development of lateritic soils. Lateritic soils are classified under soil orders Ultisols and Alfisols. They remain dry for about 4 to 5 months in an year. Kaolinite is the dominant clay mineral and gibbsite is absent in these soils.

Constraints and Management

Constraints are almost similar to that of laterite soils and hence the same management options can be followed.

4. RED SOILS

Red soils occur extensively in South and South East Asia, Western and Central Africa, Brazil, Australia, Japan, China and India. In India red soils are found in AER 7, 8, 12, 17 and 19 comprising North Eastern Hill Ranges, Himalayas, Central Highlands, Eastern and Western Ghats, Deccan Plateau and Konkan coastal lands. Red soils all together covers an area of 70 M ha in India (Sehgal, 2005) distributed mainly in the states Andhra Pradesh, Karnataka, Maharashtra, Orissa, Goa and North East region. In Kerala, red soil is found in southern parts of Thiruvananthapuram district and in pockets in catenary sequence along the foot slopes of laterite hills and mounds of Palakkad district.These soils are identified in undulating plains of lowlands with a general slope of 3 to 10%. These are mostly very deep and homogeneous in nature. The texture of the soil varied from sandy clay loam to clay loam with red to dark red colour. Gravels are rarely noticed in these soils. pH ranges from 4.8 to 5.9. These soils are nutritionally poor and possess low organic matter content. A variety of crops such as coconut, arecanut, banana, yams, pineapple, vegetables, fruit trees etc. can be grown under proper management.

In Andhra Pradesh, Tamil Nadu, Karnataka and Madhya Pradesh, red and black soils occur under similar bioclimate conditions, but on different parent materials and landforms. The red soils develop on igneous (acidic) rocks and occupy higher topographic positions whereas black soils develop on basalt (basic) rocks or on alluvium derived from basalts, occupying relatively lower topographic positions on the landscape.

Genesis

Red soils are developed in hot semi-arid to subhumid, subtropical climatic conditions, on acidic crystalline parent rocks such as granite and gneisses, occupying relatively higher topographic positions. Under such conditions, weathering is moderately intense and leads to decalcification. Some of the weathering products are leached out, leaving the resistant materials like silica, iron and alumina. Fe and Al get oxidized to form their oxides / hydroxides and impart red colour to the soil, while the clay move downwards in the percolating water and get deposited in B horizon generally forming an argillic horizon. The depleting conditions are strong enough to remove fractions of bases but not in significant amounts. Hence base saturation is moderately high ($\geq$ 35 %)

Characteristics

1. They are red to yellowish in colour due to coating of ferric oxides on ped surface with a hue of 7.5YR or redder
2. Highly variable in texture ranging from sandy to clayey
3. They are generally well drained depending on their topographic position.
4. Generally deep soils, but depth is also decided by the topographic position
5. Acidic to neutral in reaction depending on amount of iron oxides present
6. Silica:sesquoxide ratio varies from 2.5 to 3.0
7. The CEC and base saturation is low but better than that of laterite soils
8. Generally deficient in organic matter, N, P, Ca, Mg and micronutrients
9. Kaolinite, illite and chlorite clays are present in these soils, kaolinite being the dominant form

Constraints

1. Subjected to surface crusting and hardening
2. High soil erosion potential, excessive drainage and surface runoff
3. Low water holding capacity because of texture and dominance of kaolinite and illite
4. Compacted subsoils due to illuviated clay and impedes root development
5. Poor natural soil fertility
6. Soil acidity
7. Low CEC

Management

1. Addition of organic manures and incorporation of crop residues to soil to improve soil structure, water and nutrient holding capacity
2. Correct acidity through liming and other appropriate methods
3. Deep ploughing to break the subsurface hardpans
4. Follow soil test based nutritional management and in unfavourable soil situations, practice foliar nutrition
5. Good agricultural practices are to be followed
6. Undertake agronomic / engineering soil conservation measures
7. Plan the crops according to the land use plans developed for the sites
8. Adoption of integrated cropping systems could address the problems to a certain extent

5. SALT AFFECTED SOILS

Salt affected soils are that soils contain considerable amount of soluble salts and/or sodium on exchange complex. All soils invariably contain soluble salts, but under certain soil and environmental conditions excess salts accumulate in the root zone where crop growth will be adversely affected. Such soils generally occur in areas where potential evapotranspiration exceeds the precipitation ie., in arid and semiarid regions. The low annual rainfall in arid regions results salt accumulation and salinization increases with dryness. Maximum salinity is found in deserts. These soils are unproductive unless harmful salts are lessened or removed. The soluble salts generally present are the chlorides, sulphates, carbonates and bicarbonates of sodium, magnesium and calcium. Potassium salts are generally low. Under dry climates, salts of boron, nitrates and fluorides also get accumulated. In India, about 6.727 M ha is salt affected, of which 2.956 M ha is saline and the remaining 3.771 M ha is sodic (Arora *et al.*, 2016; Arora and Sharma, 2017). Major states having salt affected soils are Gujarat (2.23 M ha), Uttar Pradesh (1.37 M ha), Maharashtra (0.61 M ha), West Bengal (0.44 M ha) and Rajasthan (0.38 M ha) (Mandal *et al.*, 2018). These soils are spanned all over India covering four major ecological regions including 15 states. They are

1. Semi-arid Indo-Gangetic alluvial tract of Punjab, Haryana, UP, Delhi and parts of Bihar and West Bengal
2. Arid and semi-arid tracts of Gujarat, Rajasthan, MP and Maharashtra
3. Peninsular regions of Maharashtra, Karnataka, Tamil Nadu, AP and Odisha
4. Coastal alluvial region of AP, Odisha, Tamil Nadu, Kerala, Karnataka, Maharashtra, Gujarat and Islands of Andaman and Nicobar

Salt affected soils of India fall mainly into two categories ie., saline and sodic soils. But in certain areas with an annual rain fall of around 550 mm saline-sodic soils are also found in between saline and sodic soils as a narrow stretch (Arora and Sharma, 2017).When management options are considered, saline sodic soils are included under sodic soils (Quadir*et al.*, 2007). Saline soils are found in an area of 1.71 M ha and coastal saline soils in 1.25 M ha and sodic soils in 3.78 M ha (Mandal *et al.*, 2018). Extensive areas under sodic soils are found in Gangetic plains and saline soils in canal irrigated areas of arid and semi-arid tracts.

In states like Kerala, Odisha and West Bengal, the saline soils present are entirely different from that present in arid tracts. In these states, the acid soils predominate and the salinity is mainly due to sea water intrusion. The soils of these areas are acidic and the saline water intrusion makes it acid saline in nature. These soils remain almost wet throughout the year.

Reasons for Salt Buildup or Salinity

All soils invariably contain soluble salts, but under certain soil and environmental conditions, excess salts accumulate in the root zone which often deteriorate soil properties to such an extent that crop production is adversely affected. Development of salinity is either due to genetic reasons or anthropogenic reasons or both. They are categorized into genetic and anthropogenic reasons.

I. Genetic reasons.

1. Parent material and weathering of rocks and minerals (high in sodium)
2. Arid climate which accelerates the rates of surface evaporation and plant transpiration
3. Sea water intrusion: Sea water intrusion into coastal lands as well as into the aquifer due to over extraction and overuse of fresh water
4. Transport of salt by rivers: The salts brought down by the river water during its course of flow are spread down streams and get deposited in plains
5. Wind borne salts yielding saline fields

II. Anthropogenic reasons

1. Irrigation with poor quality water (high in carbonate and bicarbonates) and over extraction of ground water
2. Development of saline creeks due to excessive leaching
3. Seepage from canals and salt laden run-off from adjoining areas and undrained basins
4. Deposition of salts on soil surface from high subsoil water table

5. Faulty land use: Overuse of fertilizers, improper use of soil amendments, use of sewage sludge and/or treated sewage effluent, dumping of industrial brine onto the soil etc. enhances soil salinity.
6. Restricted drainage and a rising water-table

Salt affected soils are seldom formed *in situ*. Hydrological processes release salt constituents from primary and secondary minerals to surrounding water. In arid region these salts are not leached from the soil but accumulate due to high evaporation. They often occur on irrigated lands, especially in arid and semiarid regions, where annual precipitation is insufficient to meet the transpiration needs of plants. Generally adsorbed calcium and magnesium ions dominate the exchange complex of these soils, but in sodic soils sodium will be the predominant ion.

Common Sources of Salinity

1. **Rocks and minerals :** During the process of weathering of rocks, they gradually release salts. The salts primarily originate as a result of reaction like hydrolysis, hydration, carbonation, oxidation, reduction etc. Some of the common salts released are halite, gypsum, calcite, dolomite, apatite, olivine, feldspars, biotite etc. The types of salts present are decided by the kind of primary minerals and their extent of weathering. The distribution and accumulation of salts are decided by physico-geographical and geohydrological conditions.
2. **Ocean water:** Water from oceans contains "42×10^{15}" tons of dissolved salts of which 85.6 % is sodium chloride. It is the direct source of salts in coastal areas. Even rain water near the sea coast is rich in chlorides of Na and Mg.
3. **Atmospheric accession of salts:** Aerosol sprays of ocean water from atmosphere along with rain water or as a dry salt is also a common source of salts.
4. **Marine rocks and evaporites:** Marine rocks and evaporites developed as a result of geological upliftment from sea/ocean are

also important sources of salts. Eg. High salt concentration in lower Himalayas and Shivalik, Aravalli and Kutch in India.

5. **Salt springs:** Salt springs emerging from different places also contribute salts to soil since the water from interiors of earth contains large quantities of chlorides.
6. **Anthropogenic activities:** Faulty irrigation with low quality water and without proper drainage is one of the major reasons for the development of salinity in irrigated areas with low rainfall. The improper use of such poor quality waters, especially with soils having a restricted drainage, results in the capillary rise and subsequent evaporation of the soil water. This causes the development of surface and subsurface salinity, thereby reducing the value of soil resource. Such saline soils are found in command areas of several dams especially in Punjab. This salinity is also known as secondary salinity.

Origin of Salt Affected Soils

The salts get accumulated in definite zones and contribute towards soil salinization and this is mainly decided by the solubility and mobility of salts.The salt affected soils are formed as a result of following factors.

1. Continental cycles: related with movement, redistribution and accumulation of carbonates, nitrates and chlorides in inland regions
2. Marine cycles: accumulation of sodium chloride from the oceans / sea in lowland and coastal areas
3. Deltaic cycles: movement and accumulation of salts carried by rivers from the continents or delta valley ground streams or from the sea
4. Artesian cycles: salt accumulation by evaporation of deep ground water wedged up to the surface through tectonic features
5. Anthropogenic cycles: salt accumulation due to erroneous human activities

How Alkalinity Develops

The basic cations, OH^- generating anions like carbonates (CO_3^{2-}) and bicarbonates (HCO_3^-) reacts as follows.

1. **Carbon dioxide and carbonates**

$$CaCO_3 \rightarrow Ca^{2+} + CO_3^{2-}$$

$$CO_3^{2-} + H_2O \rightarrow HCO_3^- + OH^-$$

$$HCO_3^- + H_2O \rightarrow H_2CO_3 + OH^-$$

$$H_2CO_3 + H_2O \rightarrow 2H_2O + CO_2$$

2. **Role of individual cations (eg. Na^+ or Ca^{2+})**

   The particular cation associated with carbonate and bicarbonate anion influences the pH. If Na^+ is prominent in soil, more OH^- will be produced because Na is more highly water soluble than Ca^{2+}.

$$Na_2CO_3 \rightarrow 2Na^+ + CO_3^{2-}$$

3. **Role of soluble salts**

   High levels of neutral salts ($CaSO_4$, Na_2SO_4, NaCl and $CaCl_2$) lowers the pH by common ion effect. The common ion effect is usually seen as the effect on the solubility of salts and other weak electrolytes.

General Properties of Salt Affected Soils

1. **Nutrient availability:** Phosphorus is commonly deficient in these soils because it is tied up with Ca or Mg as insoluble phosphates. Availability of micronutrients such as Cu, Zn, Fe, Mn, and B are influenced by soil pH. The lower the pH they are more soluble and available. So, in alkaline conditions, plant growth may be limited by deficiencies of these metals. However, Mo is more available in alkaline conditions.
2. **Cation exchange capacity:** These soils have higher cation exchange capacity because at low pH, soil charge becomes more positive, whereas at high pH, negative charge increases.

3. **Calcium accumulation in sub surface layers:** Salt affected soils in low rainfall regions commonly have layers of $CaCO_3$, $CaSO_4$ or other such layers that can inhibit plant growth.
4. **Soil water supply:** The subsoil layers are always drier than in humid regions resulting in much competition for water by native plants. These soils therefore require greater care in water management if cultivation is a priority.

Criteria for Characterization of Salt Affected Soils

Salt affected soils are characterized mainly based on the chemical characteristics but physical characteristics also contribute to it.

A. Chemical characterization

1. Exchangeable sodium percentage (ESP)
2. Sodium absorption ratio (SAR)
3. Electrical conductivity (of 1:2.5 soil water suspension or saturation extract)
4. pH (1:2.5 soil water suspension or saturation extract)

Three primary soil properties are measured to characterize the salt affected soils

1. **Exchangeable sodium percentage (ESP):** It is the percentage of exchangeable Na ions to CEC of the soil

$$\text{ESP} = \frac{\text{Exchangeable Na (cmol kg}^{-1})}{\text{Cation exchange capacity (cmol kg}^{-1})} \times 100$$

Soils through which salt water flows often adsorb too much of Na on particle exchange sites resulting higher pH values of 8.5 to 10 and cause aggregate disintegration and dispersion. Hence ESP is considered as an important factor that decides the extent of salinization. Safer limit of ESP is <15.

2. **Sodium absorption ratio (SAR):** SAR characterizes the relation of sodium status of irrigation waters and soil solutions. This is used to estimate the amount of sodium in relation to other cations.

It gives information on comparative concentration of Na^+, to Ca^{2+} and Mg^{2+}. When SAR is 13, the soils probably will lose permeability as Ca and Mg salts are removed. Water would not be good for irrigation if its SAR $\geq$ 13.

$$SAR = \frac{Na^+}{\sqrt{\frac{Ca^{2+} + Mg^{2+}}{2}}}$$

3. **Electrical conductivity:** Salt concentration is estimated on the basis of the ability of the salt in soil solution to conduct electricity. The electrical conductivity is directly proportional to the total soluble salt concentration. Laboratory measurements of EC of the saturation extract of soil or 1:2.5 soil water suspension is practiced. It is expressed in dS m^{-1}.

   The general criteria for characterization of salt affected soils are presented in Table 4.

Table 4. Criteria for characterization of salt affected soils

Characteristics	Saline soil	Saline-sodic soil	Sodic soil
Content in soil	Excess of neutral salts	–	Excess of sodium salts
pH	<8.5	>8.5	>8.5
EC (dS m-1)	>4	>4	<4
ESP %	<15	>15	>15
Physical condition of soil	Flocculated	Flocculated	Deflocculated
Permeability to air and water	Comparable to normal soils	Moderate	Restricted
Drainage	Good	Good	Poor
Colour	White		Black
Organic matter	Slightly less than normal soils	less than normal soils	Low
SAR	<13	>13	>13

B. Physical characterization

Physical characterization is done on the basis of the following characteristics. As in the case of chemical characteristics, the values for physical characteristics are not available. The values for physical characteristics may varie with soil texture. However, the following parameters can be used for physically characterising the salt affected soils.

1. **Saturation percentage:** The maximum amount of water that a soil can hold when all the soil pores are filled with water. As the soil class move from saline to sodic saturation percentage increases because of the highly dispersed nature of sodic soils.
2. **Infiltration rate:** The infiltration rate is the velocity or speed at which water enters into the soil. For saline soils it will be near to that of normal soils but for sodic soils the infiltration rate will be very poor due to its dispersed condition. Soil dispersion hardens soil and blocks water infiltration, making it difficult for plants to establish and grow. The decreased infiltration reduced plant available water and increase runoff and soil erosion.
3. **Hydraulic conductivity:** It is the ease with which water can move through pore spaces or fractures. Saline soil possesses good permeability to air and water. The soils with well-defined structure will contain a large number of macropores, cracks, and fissures which allow for relatively rapid flow of water through the soil. When sodium-induced soil dispersion causes loss of soil structure, the hydraulic conductivity is also reduced. Sodic soil possesses very low hydraulic conductivity
4. **Particle size distribution:** It is the distribution of soil separates like sand, silt and clay. Soil texture is an important criterion to characterize saline soils. Sandy soils with 0.1 % salt concentration cause injury to the common crops while the crops grow normally in a clayey soil with same salt content.
5. **Aggregation and stability:** Soil aggregates are groups of soil particles that bind to each other more strongly than to adjacent

particles. It is an indication the soil structure. For sodic soils the aggregation will be poor while saline soils are better aggregated due to their flocculated condition as a result of more of Ca and Mg. Saline soils will be in a flocculated condition while sodic soils are deflocculated

6. **Crust formation:** Surface crusting is a characteristic of sodium affected soils. The primary causes of surface crusting are 1) physical dispersion caused by impact of raindrops or irrigation water, and 2) chemical dispersion, which depends on the ratio of salinity and sodicity of the applied water. Surface crusting due to rainfall is greatly enhanced by sodium induced clay dispersion. When clay particles disperse within soil water, they plug macropores in surface soil forming a cement like surface layer when the soil dries. The hardened upper layer, or surface crust, restricts water infiltration and plant emergence.

Damage Caused by Soil Salinity

The extent of damage caused by soil salinity ranges from very mild to severe crop loss. Shahid (2013) had summarized the damages due to increase in soil salinity. They are

1. Affects all aspects of plant development including germination, vegetative and reproductive development
2. Plants experience osmotic stress and nutritional stress resulting decline in crop yields
3. Abandonment or desertification of previously productive farm land
4. Increasing numbers of dead and dying plants
5. Increased risk of soil erosion due to loss of vegetation
6. Contamination of drinking water
7. Roads and building foundations are weakened by an accumulation of salts within the natural soil structure
8. Loss of biodiversity and ecosystem disruption
9. Lower soil biological activity due to rising saline water table

Visual Indicators of Soil Salinity and Sodicity

There are several visual indications for salinity and the effects are very clear in irrigated soils. Some of the prominent symptoms for salinity are presence of white salt crust, fluffy soil surface, patchy crop establishment, reduced or no seed germination, leaf burn, marked changes in leaf color and shape, occurrence of naturally growing halophytes etc. (Shahid and Rahman, 2011). Soil sodicity expresses the symptoms like very poor vegetative growth with variable height for plants and stunted growth, plants with shallow rooting depth, poor penetration of rain water resulting surface ponding, surface sealing and crusting etc.

Assessment of Soil Salinity and Sodicity

Visual assessment of soil salinity provides a qualitative information only. For quantitative information EC measurement of saturation extract is required. Under field condition EC measurement of saturation extract being difficult, EC of soil water suspension of ratios 1:1 or 1:2.5 or 1:5 is measured and related to the EC of saturation extract.

For sodic soils, relative level of soil sodicity can be determined in field through the use of a turbidity test on soil:water (1:5) suspensions, with the following ratings.

Table 5. Ratings for soil sodicity

Colour of the suspension	Sodicity class
Clear suspension	Non sodic
Partly turbid or cloudy	Medium sodicity
Very turbid and cloudy	High sodicity

Accurate soil sodicity diagnostics can be made by analyzing soil samples in the laboratory for exchangeable sodium percentage (ESP) and sodium adsorption ratio (SAR). Advanced methods like use of salinity probe, electromagnetic induction, salinity sensors and data loggers etc. are also available for quick determination of the extent of soil sodicity.

Classification of Salt Affected Soils

Salt affected soils are classified in to three classes viz., saline soils, saline-sodic soils and sodic soils. Sodic soils are also called as alkali soils.

1. **Saline soils:** Saline soils contain a concentration of neutral soluble salts sufficient to interfere seriously with the growth of most of the plants. The EC of saturation extract (ECe) is greater than 4 dS m^{-1}, ESP < 15, SAR > 13 and pH less than 8.5, because the salts are neutral. The chlorides and sulphates of the base forming cations are dominating in these soils. The concentration of neutral salts like chlorides and sulphates are much higher than those of alkaline counterparts like carbonates and bicarbonates. Saline soils are called **white alkali** because of the surface encrustation of salts present in white colour. But the soils are in well aggregated condition due to the dominance of calcium and magnesium salts compared to that of sodium. Under arid condition, these soils sometimes express excessive concentration of boron, fluorides and nitrates.

   These soils can be reclaimed by leaching out the salts with good quality water. If water with high SAR is used for leaching, the soil adsorbs too much Na, become very basic with a pH of 8.5 to 10.5. The soil aggregates disintegrate and disperse. These soils can become impermeable to water because the fine soil particles may seal the soil pores. Such spots could not infiltrate water and appear wet for longer periods than the adjacent areas. Because of this visual effect they are called as "**slick spots**"

2. **Saline-sodic soils or saline-alkali soils:** These soils contain appreciable quantities of neutral soluble salts and enough Na ions to seriously affect the plant growth. Here ESP > 15, EC of saturation extract > 4 dS m^{-1} and pH > 8.5 because of the presence of neutral salts. The SAR is at least 13 in these soils.

   Leaching of saline sodic soils with poor quality water will increase pH unless Ca and Mg salt concentration is high in soil or in the irrigation water. Increase in pH occurs because once neutral salts are removed, the exchangeable Na readily hydrolyse and release OH ion, causing a sharp increase in pH. The high ESP badly affect the physical properties of soil by dispersing the clay particles and make the soil poorly structured and in due course it is converted to sodic soils.

3. **Sodic soils or Alkali soils:** Sodic soils contain very small quantities of soluble salts but are rich in Na salts. Carbonates and bicarbonates of sodium are the dominant salts. Gypsum is nearly absent. In these soils, ESP >15, EC of saturation extract < 4 dS m^{-1} and pH > 8.5 and SAR > 13 due to the dominance of sodium. The high values of ESP adversely affect the physical properties of soil. High sodium saturation cause dispersion of clay particles making the soil poorly structured. The dispersed clay particles move downwards through soil pores and form a dense or compact layer of very low permeability at the lower layers of soil. Hydraulic conductivity of sodic soils is very low. Hence when irrigation is given, the water stagnates at the surface due to the clogging of soil pores by the dispersed clay particles, which on drying results crust formation.

   The detrimental effects on plant growth are not only due to the toxicity of Na^+, $HCO3^-$ and OH^- ions but also due to reduced water infiltration and aeration. The high pH is largely due to hydrolysis of sodium carbonate. The Na on exchange complex also undergoes hydrolysis.

   Irrigating saline or saline-sodic soils with poor quality water (low in Ca and Mg) will result in the formation of sodic soils. Due to the deflocculating influence of Na, sodic soils have very poor physical characteristics. Leaching of these soils with poor quality water will remove all the neutral salts and Na become more dominant. Because of the extreme alkalinity due to Na, the surface of these soils is usually discoloured by the dispersed humus carried upward by capillary water and deposited at the surface when it evaporates. Hence these soils are called as **black alkali**.

Table 6. Comparison of properties of salt affected soils

Property	Saline soil	Saline-sodic soil	Sodic soil
pH	< 8.5	> 8.5	>8.5
EC of saturation extract (dS/m)	> 4	>4	<4
SAR	<13	At least 13	>13
ESP	<15	>15	>15

Salt Affected Soils and Plant Growth

Salt accumulation in soil affects plant growth in different ways. The sensitive crops find it difficult to thrive in soils with an EC > 2 dS m^{-1} while halophytes can easily overcome very high EC values. Salt resistant / tolerant varieties of crop plants are able to grow under high EC values in a range of 4-8 dS m^{-1}.

The sensitive, resistant or tolerant plants show differential response to salinity. The general response is through osmotic effect, though specific ion effect is also prevalent. High salt concentration increases the potential forces that hold water in the soils and makes it more difficult for plant roots to extract the moisture (Exosmosis). Salt in the soil solution forces a plant to exert more energy to absorb water and to exclude salt from metabolically active sites. The saltier the soil, the wetter it must be kept to dilute the salts.The specific ion toxicity due to high concentration of sodium, chloride or other ions that can occur.

Effects of High Salt Concentration

1. Inhibit or slows down seed germination
2. Adversely affects the growth of plants. Plants vary in their salt tolerance and are classified into four groups based on their tolerance to salinity.

Table 7. Classification of plants based on their tolerance to salinity

Tolerant	Moderately tolerant	Moderately sensitive	Sensitive
Sugar beet	Wheat	Rice	Lemon
Date	Sorghum	Pea	Onion
Cotton		Cabbage	Tomato

3. Toxicity due to specific ions
    1. Soils with ESP > 15, Na disperses the soil and form hardened crusts. These hardened crusts adversely affect plant growth by preventing root penetration

2. Caustic influence of high pH due to sodium carbonate and bicarbonate
3. Toxicity of bicarbonate and other anions
4. Adverse effects of active Na
5. Low nutrient availability under high pH
6. Breakdown of structure which cause oxygen depletion, poor infiltration and percolation
7. High chloride content in association with Na causes leaf damage

Adaptation Mechanisms

Plants have their own mechanisms to tolerate or resist excess salt concentration in soil. They accomplished it mainly through the following three mechanisms.

1. Tolerance to osmotic stress
2. Na^+ or Cl^- exclusion
3. Tolerance of tissues to accumulated Na^+ or Cl^-

1. **Tolerance to osmotic stress**: The osmotic stress immediately reduces cell expansion in root tips and young leaves, and cause stomatal closure. It varies with plant species and external conditions like salt content, water availability, humidity, transpiration rate, leaf water potential etc. The tolerant species are known as halophytes and others as glycophytes. Plant growth responds to salt stress in two phases.
   a) Rapid osmotic phase where growth of young leaves is inhibited due to decrease in ability of plant to take up water. Plants enact mechanisms to mitigate osmotic stress by reducing water loss while maximizing water uptake.
   b) Slower ionic phase where the senescence of mature leaves is speeded up. Initial osmotic stress is followed by the accumulation of ionic Na^+ with additional Cl^- stress, which results in early senescence of mature leaves.

2. **Na^+ ions exclusion from leaf blades**: Na exclusion ensures that Na does not accumulate to toxic concentrations within the leaves. A failure in Na^+ expulsion, cause premature death of plants.
3. **Tissue tolerance to accumulated Na^+ ions**: Requires compartmentalization (usually stored in vacuoles) within leaves to avoid excess concentrations within cytoplasm, especially in mesophyll cells. Plants allow more Na to get accumulated in older leaves. Certain plants absorb large quantities of salt and accumulate it in vacuoles of leaves for osmotic adjustment.

Reclamation and Management of Salt Affected Soils

There are two approaches to tackle the problem of soil salinity / sodicity. One is to reclaim salt affected soils and the second is to manage them using suitable agricultural options such as cultivation of salt resistant crops, saline aquaculture, mangrove forests etc. Not all salt affected soils can be reclaimed practically and economically but can be managed. However, gypsum based sodic soil reclamation, improved agroforestry techniques, cultivation of salt resistant crops etc. are some of the all-time popular technologies for addressing the salinity / sodicity problems. Wherever the reclamation of salt affected soils are attempted, two conditions have to be taken care of viz., salt balance and leaching requirement of soil.

Salt Balance

The relationship between the quantity of salt brought into an area with irrigation water and the quantity of salt removed in the drainage water has been referred to by Scofield (1940) as salt balance of the area. If favourable salt balance occurs, the outgoing salt (drainage) should be equal or more to that of incoming salt (irrigation). Proper water management and good quality water could maintain a salt balance in salt affected areas.

Leaching Requirement

Since salts are continuously added through irrigation water in salt affected areas, additional leaching must be provided to remove the salts contributed by the irrigation water. For this, additional quantity of irrigation water has to be given. This additional water needed for leaching, over that needed to wet the profile is called leaching requirement. Leaching Requirement is defined as the fraction of irrigation water that must be leached through root zone to keep the salinity of soil below specific limit. Plant sensitivity to salt tolerance is also to be taken into account.

$$LR = \frac{EC_{iw}}{EC_{dw}}$$

EC_{iw} – EC of irrigation water

EC_{dw} – EC of drainage water

Reclamation Methods

The salt affected soils have to be reclaimed for making them productive. The process of accumulation of salts and buildup of sodium have to be reversed. Provision of drainage, replacement of Na from exchange complex and leaching out below root zone have to be ensured for proper crop growth.

Reclamation includes methods for either temporary removal or permanent removal of salts from soil. The permanent reclamation is very difficult. Reclamation of soil on temporary basis can be done by

1. Removing salt crust from soil surface
2. Ploughing surface salt crust to deeper levels
3. Neutralization effect of salts

However, reclamation for longer period is possible by

1. Lowering of water table, if high
2. Improving infiltration rate
3. Leaching of salts in saline soils

4. Replacing excessive Na by Ca and remove replaced products
5. Suitable management practices

The reclamation methods can be classified into four classes based on the type of method selected.

1. Physical methods for amelioration
2. Hydro-technical methods for amelioration
3. Chemical methods for amelioration
4. Biological methods for amelioration

I. Physical methods for amelioration

The commonly followed methods are

1. Deep ploughing: Helps to break any impervious layer present and improves internal drainage which will facilitate transportation of salts to deeper layers
2. Sub-soiling : Breaking of hardpan or cemented sub-soil layer occurring at various depths to improve internal drainage and facilitate transportation of salts downwards
3. Sanding : Application of sand and thorough mixing is done to bring permanent changes in soil texture, and improve soil permeability and air-water relations in the root zone. For sodic soils special care should be taken to see that sufficient quantity of sand has been added so that soil cementing will not occur.
4. Profile inversion: Practiced under conditions whereas surface soil is good and at deeper layers it is saline or sodic.
5. Scarping of salts from the surface: Scraping of salts from surface for few centimeters is a temporary measure for improving plant growth.

II. Hydro-technical methods for amelioration

Salt leaching with ponded fresh water, sub-surface drainage, mulching between two irrigations and during fallow period, irrigation management etc. are some of the effective hydro technological interventions to address the problems of water logging and salinity (Arora and Sharma, 2017). Hydro technological methods are highly effective for the reclamation and management of saline soils. Sub-surface technology had increased the crop yields for paddy, wheat and cotton (Sharma *et al.*, 2015).

Hydro technological methods involve basically the removal or displacement sodium ions from the exchange complex of sodic soils by leaching with good quality water. Here sufficient water must pass through the soil to decrease the salt content below the safe limit and maintain proper salt and water balance. While this method is adopted make sure that

- Sufficient quantities of good quality irrigation water is passed through the soil to decrease the salt concentration below the permissible limit and to maintain proper salt balance
- Provision for adequate internal and surface drainage are available and
- Suitable areas should be available for salt disposal, facilities for evaporation ponds can be considered for safe disposal of drained water.

The extent of leaching required to remove salts depends on

1. Soil properties (texture, porosity, structure)
2. Initial salinity and chemical composition of salts
3. Depth of water table
4. How deep salts are to be washed
5. Percentage of salts to be removed
6. Methods of leaching (drip and sprinkler methods are more effective than ponding)
7. Drainage system: Subsoil drainage system or surface drainage. To be selected depending on the feasibility, effectiveness and cost.

III. Chemical methods for amelioration

Before going for chemical technologies, make sure the availability and feasibility of other practices like land levelling, bunding, drainage for removal of excess water, availability of good quality water, application of amendments, selection of crops and efficient nutrient management.The amount and type of amendments required depends on soil pH, EC and ESP. Soluble Ca sources are recommended for calcareous soils, while acids or acid formers are recommended for calcareous soils. Gypsum followed by pyrite are the most popular chemical amendments.

Reclamation of sodic soils required neutralization of alkalinity and replacement of most of the Na in the exchange complex by Ca. This can be achieved by application of chemical amendments followed by leaching to remove soluble salts and other reaction products. The type and the quantity of chemical compounds required depends on the soil properties and desired rate of Na replacement. The chemical amendments generally used are mainly sulphates of Ca, Fe and Al.

1. **Soluble sources of Ca:** Gypsum ($CaSO_4 \cdot 2H_2O$), calcium chloride ($CaCl_2$) and phosphogypsum (an industrial byproduct from P fertilizer plant)
2. **Acids or acid formers:** Elemental sulphur, iron sulphate, aluminium sulphate, pyrite and lime-sulphur (used in soils containing sufficient quantities of $CaCO_3$ which may get solubilised in the presence of acids)

   When gypsum is applied to sodic soil the reaction is as follows

$$Na_2CO_3 + CaSO_4 \rightleftharpoons CaCO_3 + Na_2SO_4$$

$$\text{Clay-2Na} + CaSO_4 \rightleftharpoons \text{Clay–Ca} + Na_2SO_4$$

Sulphur reacts with oxygen and water to form sulfuric acid which will again react with sodium carbonate to form sodium sulphate which can be leached out from soil.

$$2S + 3O_2 \rightleftharpoons 2SO_3$$

$$SO_3 + H_2O \rightleftharpoons H_2SO_4$$

$$H_2SO_4 + Na_2CO_3 \rightleftharpoons CO_2 + H_2O + Na_2SO_4$$

Similarly, it can react with $CaCO_3$ and clay.

3. **Organic sources:** Farm yard manure and industrial byproducts like phosphogypsum, pressmud, molasses, acid wash, effluents from milk plants etc. are helpful for reclamation of sodic soils. They can be used to alleviate problems with excess salts without irrigation. Addition of organic amendments along with gypsum reduces the gypsum requirement (Arora and Sharma, 2017).

Gypsum requirement (GR)

It is the amount of gypsum to be added to a soil to reduce the exchangeable sodium percentage (ESP) to 10

$$GR = \frac{[ESP_{initial} - ESP_{final}] \times CEC}{100}$$

Eg. GR of a soil having an initial ESP of 60 and final ESP 10, CEC-30 cmol kg^{-1} is

Milli equivalent (Me) of Ca^{2+} per 100 g of soil $= \frac{[60-10]\times 30}{100} = 15$

One ha furrow slice soil (15 cm depth) $= 2 \times 10^6$ kg

One me of replaceable gypsum = 860 mg kg^{-1} of amendment

Amount of gypsum required $= 860 \times 2 \times 10^6 \times 15 = 25800$ kg

GR depends on the exchangeable Na to be replaced, exchange efficiency and depth of soil to be reclaimed. Gypsum is the most economical and commonly used chemical amendment. One tone of gypsum is equivalent to "0.18 t sulphur = 0.75 t lime sulphur = 1.62 t iron sulphate".

Lab estimation is done by equilibrating a sodic soil with gypsum solution of known Ca concentration and then estimating the Ca deficit in the extracted solution. This determination includes the Ca^{2+} required to replace Na^{+} and to neutralize the alkalinity. The method is described below.

A saturated solution of gypsum is prepared by dissolving 5 g gypsum in one liter of water. Shake it for 10 min, and filter it. Five g of soil is taken and 100 ml of the above extract is added. Shake for half an hour. Filter it and find out "Ca+Mg" content.

Calculation:

Concentration of Ca in the saturated Gypsum solution $= C_1$ cmol kg^{-1}

Conce ntration of "Ca+Mg" in the filtrate $= C_2$ cmol kg^{-1}

$$\text{Gypsum requirement in meL}^{-1}\ 100\ \text{g} = \frac{(C_1 - C_2) \times 100 \times 100}{1000 \times 5}$$

Table 8. Estimated efficiency of various soil amendments

Amendment	Amount (t) equivalent to 1 t gypsum
Gypsum ($CaSO_4{\cdot}2H_2O$)	1.00
Sulphuric acid	0.57
Sulphur	0.18
Iron sulphate	1.62
Lime sulphur (Ca polysulphide with 24% S)	0.75

IV. Biological methods for amelioration

Several biological methods are also adopted to overcome the ill effects of salinity but their use alone could not correct all the problems associated with salt accumulation. In most of the cases it can supplement the chemical methods and provide a better soil environment for promoting crop growth. The most common practice is application of organic manures to soil which will improve soil physical conditions and salinity amelioration to a certain extent. Some of the practices generally followed are

- Addition of organic manures to soil which will improve action of plant roots and biological activity of soil
- Application or incorporation of green manures where their decomposition increase CO_2 and organic acid content, which will mobilize Ca by dissolving it
- Application of coir pith
- Application of industrial byproducts like press mud
- Growing of crops on problem soils and/or their incorporation at the stage of maximum biomass productivity
- Growing of salt tolerant crops
- Ponding of water which decreases the soil pH in alkaline soils

V. Associate practices to improve reclamation efficiency of sodic soils

1. **High salt water dilution technique:** A significant innovation in reclaiming sodic soil and saline – sodic soils is the initial leaching with salt water (Paradoxical situation). But a high salt content of a Ca and Mg in water keeps sodic soils flocculated. The floccules have large pores between them as do aggregates and allow penetration of leaching water. Initially the soil is washed with moderately salty water and after removing much of Na, low salt water is used. For final leaching good quality water with low salt content is used.
2. **Synthetic conditioners:** Used to improve physical properties and facilitate better drainage: Eg. polyacrylamide polymers used along with gypsum @ 50 mg kg^{-1} improved the drainage.
3. **Electro-amelioration:** It aims the improvement of physical and chemical properties of soil by applying electric current (55 V direct current was maintained in soil, and current densities ranged from 0.04 to 0.35 mA cm^{-2}). This enhances the release of Ca from

precipitate carbonates, which subsequently displaces the Na from the soil exchange complex, which can be easily leached out. This has improved soil physical properties. Practically this method is not that much popular.

4. **Salt precipitation:** Elimination of salts and exchange Na from soils by leaching has been one of the satisfactory methods, but the leached salts are washed into ground waters or streams, making them saltier. Hence a new concept – precipitation of salts. Here instead of leaching salts completely away they are leached to 0.9 – 1.8 m deep, where much of the salt would form slightly soluble gypsum or carbonates during dry cycles and not react any longer as soluble salts.

5. **Phytoremediation / bioremediation:** Phytoremediation of salt affected soil refers to the process of removal of salts from soil by growing salt extracting plants. They absorb salts from soils and accumulate in their biomass and thus decreases Na concentration in soil and also improve soil physical, chemical and biological properties. Plants like *Sesbania aculeata* and *Leptochloa fusca* possess phytoremediation ability for salts (Qadir *et al.*, 2007). Reclamation of sodic and saline soils through afforestation is an approved practice and some of popular species used are *Prosopis juliflora*, *Acacia nilotica*, *Casuarina equisetifolia* etc. (Dagar and Minhas, 2016). Many of the grass species are also useful for sodic soil reclamation.

   The bio-remediation approach, which involves plant-microbial interaction can also be exploited for sodic soil reclamation. Halophytic bacteria have the potential to remove sodium ions from soil and increase metabolic and enzymatic activities in plants and can be used for remediation of sodic and saline soils (Arora *et al.*, 2016).

6. **Saline aquaculture:** Inland saline aquaculture is another option for the management of degraded sodic soils. It can be profitably put to shrimp culture. In Kerala, the popular *Pokkali* cultivation consists of one rice with salt resistant varieties followed by shrimp culture is successfully being practised. The system is purely organic and did

not need any chemical fertilisers. In coastal areas of Andhra Pradesh, many farmers have converted their rice fields to high remunerative aquaculture (Singh, 2009). But such practices should not adversely affect the rice area and production.

Disposal of excess salts

It is the major problem in salt affected areas. Certain countries had amended law and regulation that water released back to ground water or surface water should be with <500 mg L^{-1} salts. Construction of evaporation ponds and letting drained water to these ponds, so that salts precipitated on drying can be safely removed. Techniques used to reduce salt contents are

1. Use of reverse osmosis membrane system
2. Good quality water is added to dilute poor quality water

Management of Salt Affected Soils

It is not always possible to eliminate salts, completely; hence management of such soils only can be done to minimize salt damage. The practices that can be followed are

1. Maintain high water content in soil, near field capacity. This will dilute salts and reduce its toxic effects.
2. Limited leaching before planting to move down the salts and periodic leaching during crop period.
3. Adjust the planting position to avoid salt accumulation zones
4. Choice of crops – decided by tolerance of crops, value of the crop and adaptability to climate.

Crop growth and electrical conductivity are highly related and the EC which can be tolerated by plants are presented below.

Table 9. Ratings for electrical conductivity related to plant growth

EC of saturation extract (dS m^{-1})	Plant growth condition
<2	Salinity effects negligible
2-4	Yield of very sensitive crops restricted
4-8	Yield of many crops restricted
8-16	Only tolerant plants
>16	Only very tolerant plants survive

Three kinds of management practices are generally followed.

1. **Eradication:** By drainage or by leaching salts are removed from soil. This is highly suited for saline soils.
2. **Conversion:** By the use of amendments like gypsum, elemental sulphur, pyrites, iron sulphate, aluminium sulphate etc. the Na^+ ions are replaced by Ca^{2+} ions. This is highly suited for sodic soils.

$$2NaHCO_3 + CaSO_4 \rightleftharpoons CaCO_3 + CO_2 + H_2O + Na_2SO_4$$

$$Na_2CO_3 + CaSO_4 \rightleftharpoons CaCO_3 + Na_2SO_4$$

3. **Control:** It includes practices like retardation of evaporation by mulching and use of salt resistant crops.

CHAPTER 5

SOILS WITH BIOLOGICAL PROBLEMS

Biological activity in soil is fundamental for plant growth and is an indicator of soil's life supporting ability. The nutrient cycling and the deposition of organic carbon to soil are decided by the biological activity. It regulates the enzyme activity and release of metabolites into the soil by solubilizing essential mineral nutrients, promote the decomposition of plant and animal remains and catalyze the degradation of xenobiotics. An increase in soil organic matter leads to greater biological activity and diversity in the soil.

Biological problems in soil often results from unhealthy management practices and anthropogenic influence. Soil organic carbon (SOC) is the main source of energy for soil microorganisms and a trigger for nutrient availability through mineralization. An imbalance in the above results soils with biological problems.

Soils with biological problems face mainly two kinds of risks.

1. Low organic matter content of the soil and
2. Low biological activity of soils.

1. SOILS WITH LOW ORGANIC MATTER

The reasons for the formation of low organic matter soils may vary from the inherent factors such as climate and soil texture to anthropogenic reasons like intense cultivation without addition of organic matter. The warm humid tropical climate favours organic matter decomposition and because of that, tropical soils are always lower in organic matter content compared to cool dry temperate soils. Similarly organic matter decomposes faster in well aerated soils and is much slower in saturated wet soils.The soils with an organic carbon content of <0.5 % is rated as soils with low organic matter.

Soil organic matter (SOM) is the organic component of soil, consisting of three primary parts: including small (fresh) plant residues and small living soil organisms, decomposing (active) organic matter and stable organic matter (humus). SOM serves as a reservoir of nutrients for crops, provides soil aggregation, increases nutrient exchange, retains soil moisture, reduces compaction, reduces surface crusting, and increases water infiltration into the soil. Soils with low SOM can be found in any part of the world due to various reasons, but more prominent in sandy and arid tracts.

The primary source of SOM is the vegetal detritus. The organic compounds added to the soil through vegetal detritus include carbohydrate, fat, lignin, protein and lipid. The members of higher trophic levels contribute to soil organic matter through their metabolic products and even corpses. These organisms that consume vegetable matter and the residue of which are then passed to the soil are the secondary sources of SOM. Charcoal formed by incomplete burning of biomass, which is resistant to decomposition added to soil also contributes towards SOC pools.

The vegetal detritus is being decomposed by soil microbes through enzymatic biochemical processes, obtain the necessary energy from the same matter, and release mineral compounds that plant roots can take up. As vegetal detritus decomposition takes place, a microbially resistant product "humus"comprising humic acid, fulvic acid and humin is formed. Secondly, some new compounds like polysaccharides and polyuronides are synthesized. These compounds are the basis of humus. New reactions occur between these compounds and proteins or other nitrogen containing products to form compounds that incorporate nitrogen and thus mineralization is avoided. Other nutrients are also protected in this way from mineralization.

Functions of Soil Organic Matter

Organic matter exists on the soil surface as raw plant residues which help to protect the soil from the effect of rainfall, wind and sun. Organic matter within the soil helps to improve moisture retention, water holding capacity, soil structure, and maintenance of tilth and minimize erosion. It is the major reserve of nutrients. As soil organic matter is derived mainly from plant residues, it contains all the essential plant nutrients. On decomposition of organic matter, nutrients are released to soil.

Soil organic matter plays a major role in improving soil structure. The active and some of the resistant soil organic components, together with microorganisms especially fungi are involved in binding soil particles into larger aggregates and thereby improves soil structure, aeration, water infiltration and resistance to erosion and crusting.

In order to maintain the soil carbon and nutrient cycling systems, a dynamic equilibrium between the gains and losses of soil organic matter has to be maintained. The rate of organic matter addition should be able to meet the loss due to decomposition, plant uptake and other losses. If the rate of addition is higher than the rate of decomposition, soil organic matter buildsup in soil or vice versa. In soil, normally the pool of organic carbon exists in dynamic equilibrium between gains and losses. Soil may therefore serve as either a sink or source of carbon, through carbon sequestration or greenhouse gas emission respectively, depending on exogenous factors.

Reasons for Decline in SOM

Variations in SOM content over the long-term basis are mainly because of climatic, geological and soil forming factors while for short-term periods, vegetation disturbances and land use changes are the main reasons. The factors responsible for SOM decline are

1. **Climate:** In natural ecosystems, climate is the main driver of change in SOM through the effects of temperature, moisture and

solar radiation. SOM increases with precipitation and reduces with temperature.

2. **Vegetation type:** Crop variety, species management, residue management, nature of farming etc. influence the SOC content. Deforestation, biomass burning, intensive farming etc. causes a decline in SOM content
3. **Soil properties:** Soil type is a major factor involved in the stabilization mechanisms of SOM by means of physical preservation. Tillage and soil disturbance have an adverse effect on SOM build up.
4. **Human activities:** Land use and management had a critical role on maintenance of SOM. Management systems affect SOM mainly through:
   a) the input rates of organic matter and its decomposability
   b) the distribution of photosynthates in roots and shoots and
   c) by the physical protection of SOM.

The SOM cycle is affected by other external drivers and pressures, such as government policies, technological developments, climate change and demographic trends.

Constraints

Major constraints are

- Reduced microbial biomass activity
- Reduced nutrient mineralization due to a shortage of energy sources
- Poor soil physical properties like low aggregate stability, infiltration, drainage and airflow
- Less diversity in soil biota with a risk of the food chain equilibrium and disturbance in the soil environment
- Poor crop growth

Management of Soils with Low SOM

1. **Addition of organic inputs:** Soil organic matter generally increases in soils that receive frequent additions of organic materials and where biomass production is high. Plant residue with a narrow C:N ratio decomposes more quickly than those with a wide C:N ratio and do not increase soil organic matter levels as quickly.
2. **Resort to minimum tillage:** Excessive tillage destroys soil aggregates, increasing the rate of soil organic matter decomposition. Stable soil aggregates increase active organic matter and protect it from rapid microbial decomposition. Reduce or eliminate tillage that causes a flush of microbial action that speeds up organic matter decomposition and increases erosion. Measures that increase soil moisture (up to field capacity), soil temperature and optimal aeration accelerate SOM decomposition.
3. **Selection of cropping systems:** Resort to cropping systems that follows continuous no-till / minimum till involving legumes, cover crops, diverse rotations with high residue crops and perennial legumes or grass used in rotation.
4. **Erosion preventive measures:** Reduce soil erosion using appropriate measures. Most of the SOM is in the topsoil. When soil erodes, organic matter goes with it. Saving soil and SOM go hand in hand.
5. **Scientific and judicious fertilizer application:** Soil test based fertilization may be followed to maintain nutrient balance within the soil. Proper fertilization encourages growth of plants, which increases root and top growth. Increased root growth can help to build or maintain SOM, even if the top growth is removed.
6. **Use of perennial forages:** Growing perennial grasses allows annual die back and regrowth. Their extensive root systems and aftermath contributing organic matter to soil favours carbon sequestration and thereby increases soil organic matter content. Fibrous root systems of perennial grasses are particularly effective as a binding agent in soil aggregation.

2. SOILS WITH LOW BIOLOGICAL ACTIVITY

Soil biology plays an active role in all the essential functions of soil. Soil food web, the biological component of soil regulates all these functions. Biological problems arise in soil mainly due to the mismanagement which will either degrade the microbial habitat or remove food resources or kill soil organisms. The poor performance of soil microorganisms adversely affects the soil productivity. Although management practices are known to impact soil biology, there is limited knowledge to support the development of detailed management strategies so as to maintain a healthy food web in soil. Soil organisms play a vital role in the sustainable functioning of ecosystems.

Functions of the Soil Food Web

1. **Energy transfer:** The energy cycle begins when the sun's energy is captured by the plants which are finally acted upon by the detritus-based (below ground) food web. Along with the energy transfer, the nutrients are released to soil. Nutrient availability is governed by the detritus based food web.
2. **Nutrient cycling:** Soil biota regulates the flow and storage of nutrients by decomposition of plant and animal residues, N fixation, nutrient transformations and release etc. Even the applied fertilizers may pass through soil organisms before being utilized by crops.
3. **Soil stability and erosion:** Soil organisms play an important role in forming and stabilizing soil structure. Fungal filaments and exudates from microbes and earthworms bind soil particles together into stable aggregates that improve water infiltration and protect soil from erosion, crusting and compaction.
4. **Water quality and quantity:** By improving or stabilizing soil structure, soil organisms help to reduce runoff, and improve the infiltration and filtering capacity of soil. In a healthy soil ecosystem, soil organisms reduce the impacts of pollution by buffering, detoxifying, and decomposing potential pollutants. Bacteria and other microbes are increasingly used for bioremediation of contaminated water and soil.

5. **Plant health:** A relatively small number of soil organisms cause plant disease. A healthy soil ecosystem has a diverse soil food web that keeps pest organisms in check through competition and predation. Some soil organisms release compounds that enhance plant growth or reduce disease susceptibility. Plants may exude specific substances that attract beneficial organisms or repel harmful ones, especially when they are under stress, such as grazing.

Components of Soil Food Web

Soil food web is composed of decomposers, bacteria, actinomycetes (filamentous bacteria), saprophytic fungi, grazers and predators, litter transformers and mutualists.

1. **Decomposers, actinomycetes and saprophytic fungi:** They degrade plant and animal residue, organic compounds, and some pesticides. Bacteria generally, but not exclusively, degrade the more readily decomposed (narrow C:N ratio) materials, compared to fungi, which can use more chemically complex materials.
2. **Grazers and predators:** Protozoa, mites, nematodes, and other organisms "graze" on bacteria or fungi; prey on other species of protozoa and nematodes; or both graze and prey. Grazers and predators release plant-available nutrients as they consume microbes.
3. **Litter transformers:** This group consists of arthropods which are invertebrates with jointed legs, including insects, spiders, mites, springtails, centipedes and millipedes. Many soil arthropods shred and consume plant litter and other organic matter, increasing the surface area accessible to decomposers. The organic matter in their fecal pellets is frequently more physically and chemically accessible to microbes than was the original litter. Some litter transformers, especially ants, termites, scarab beetles, and earthworms, are "ecosystem engineers" that physically change the soil habitat for other organisms by chewing and burrowing through the soil. Microbes (decomposers) living within their guts break down the plant residue, dung, and fecal pellets consumed along with the soil.

4. **Mutualists:** Mycorrhizal fungi, nitrogen fixing bacteria, and some free-living microbes have co-evolved together with plants to form mutually beneficial associations with plants. Mycorrhizae are associations between fungi and plant roots in which the fungus supplies nutrients and perhaps water to the plant, and the plant supplies food to the fungi. These fungi can exist inside (endomycorrhiza) or outside (ectomycorrhiza) the plant root cell wall.
5. **Pathogens, parasites and root feeders:** Microorganisms that cause disease make up a tiny fraction of the organisms in the soil, but have been most studied by researchers. Disease causing organisms include certain species of bacteria, fungi, protozoa, nematodes, insects, and mites.

Soil Biological Activity

Soil biological activity is determined by factors at three different levels.

1. **At the scale of individual organisms:** It is determined by the soil atmospheric parameters such as temperature and moisture, and pH and EC of the microbial habitats.
2. **At the scale of populations:** The amount of habitat diversity, types of habitat disturbances, diversity and interactions among various soil populations, and availability of food, air and water decide the activity.
3. **At the scale of biological processes:** Functions like nutrient cycling and release are affected by the interaction of biological population with physical and chemical soil properties. The management practices that will weaken soil structure over time and that reduce the surface residues decreases soil stability and alter the microhabitats. Monocropping is such a practice that decreases biological diversity and activity of soil, leading to biological sickness.

Indicators of Soil Biological Activity

Soil biological activity which is an indication of soil's life supporting ability is further indicated by the following parameters.

1. **Presence of earthworms:** The role of earthworms in maintaining biological health of soil is remarkable. It plays a key role in modifying the physical structure of soils by producing new aggregates and pores, which improves soil tilth, aeration, infiltration and drainage. Earthworms produce binding agents responsible for the formation of water-stable macro-aggregates. They improve soil porosity by burrowing and mixing soil. As they feed, earthworms participate in plant residue decomposition, nutrient cycling and redistribution of nutrients in the soil profile. Their casts, as well as dead or decaying earthworms, are a source of nutrients. Roots often follow earthworm burrows and take up available nutrients associated with casts.

   Low or absence of earthworm population is an indication of little or no organic residues in the soil and / or high soil temperature and low soil moisture.

2. **Soil respiration rate:** Another indication for biological activity is soil respiration rate. Carbon dioxide release from the soil surface is referred to as soil respiration. The CO_2 release is due to microbial respiration, plant root and faunal respiration, and eventually from the dissolution of carbonates in soil solution. Soil respiration reflects the capacity of soil to support soil life including crops, soil animals and microorganisms. Reduced soil respiration rates indicate that there is little or no microbial activity in the soil. With reduced soil respiration, nutrients are not released from SOM to feed plants and soil organisms.

3. **Soil enzyme activity:** Yet another factor that indicates soil biological activity is soil enzymes. Soil enzymes control the rate of organic matter decomposition and release of plant available nutrients. Enzymes are specific to a substrate and have active sites that bind with the substrate to form a temporary complex. The enzymatic reaction releases a product, which can be a nutrient contained in the substrate.

   Absence or suppression of soil enzymes prevents or reduces processes that can affect plant nutrition. Poor enzyme activity can result in an accumulation of chemicals that are harmful to the environment; some of these chemicals may further inhibit soil enzyme activity.

Constraints

1. Very low rate of organic matter decomposition
2. Low nutrient release and availability
3. Poor plant growth
4. Biological sickness due to low community diversity, nutrient cycling, and presence of toxic compounds
5. Loss of soil health

General Management Strategies

Management plans for improving the diversity and functioning of a soil biological community should consider the following aspects while formulating a strategy.

- Timing of management practices and disturbances
- Duration and degree of their effects on soil biology
- Seasonal influences on management and disturbances and
- Regenerating capacity of the soil community

The management strategies that can be adopted for improving biological activity are

1. **Manage soil organic matter:** Regular inputs of organic matter are essential for supplying the energy that drives the soil food web. Each source of organic matter favors a different mix of organisms. Thus, a variety of sources may support a variety of organisms. The location of the organic matter: whether at the surface, mixed into the soil, or as roots also affects the type of organisms that dominate in the food web.
2. **Improve plant productivity:** Under any land use, organic matter inputs to the soil can be increased by improving plant productivity and increasing annual biomass production. In particular, good root growth is important for building soil organic matter. High biomass production should be combined with other organic matter management

practices including minimizing residue removal and tillage, growing cover crops, and adding manure, mulch, or other amendments.

3. **Manage for diversity:** The diversity of plant assemblages across the landscape and over time promotes a variety of microbial habitats and soil organisms. Up to a point, soil biological function generally improves when the complexity or diversity of the soil biological community increases. Many types of diversity should be considered, such as diversity of land uses, root structures and soil pore sizes. Diversity is desirable over time as well as across the landscape. Diversity over time can be achieved with crop rotations. Rotated crops put a different food source into the soil each year, encouraging a wide variety of organisms and preventing the build-up of a single pest species.
4. **Keep the ground cover:** Ground cover at or near the surface moderates soil temperature and moisture, provides food and habitat for fungi, bacteria, and arthropods, and prevents the destruction of microbial habitat by erosion. Minimize the length of time each year that soil is bare by maintaining a cover of living plants, biological crusts, or plant residue at the surface. Living plants are especially important as cover because they create the rhizosphere. Maintaining a rhizosphere environment is one of the important benefits of using cover crops. In addition to preserving microbial habitat, cover crops help build and maintain populations and diversity of arthropods by preserving their habitat for an extended portion of the growing season.
5. **Manage disturbances:** Some soil perturbations are a normal part of soil processes, or are a necessary part of agriculture and other land uses. However, some disturbances significantly impact soil biology and can be minimized to reduce their negative effects. These disturbances include compaction, erosion, soil displacement, tillage, catastrophic fires, certain pesticide applications, and excessive pesticide usage.
6. **Improving earthworm population:** The practices that boost earthworm population viz., tillage management (no-till, strip till, ridge

till), crop rotation (with legumes) and cover crops, manure and organic by-product application etc. improve soil biological activity. Regulation soil pH and drainage are also helpful in this regard.

CHAPTER 6

SOILS WITH PROBLEMS DUE TO ANTHROPOGENIC REASONS (DEGRADED SOILS)

From the time immemorial, the association of humans with soil is well known. Once the equilibrium between the two had lost, soil degradation will sets in. Soil degradation is a human-induced or natural process which impairs the capacity of soil to function. The origin, flourishing and existence of world civilizations are very much related to soil. As an example, the collapse of Mesopotamian civilization in 2200 BC was due to faulty irrigation in dry climate which caused salinization and consequent failure of agriculture.

Entire world faces a resource crunch during the late eighties and early nineties after the green revolution era. There was a decline or stagnation in soil and crop productivity as evidenced by the failure of ensuring sustainability in agriculture sector. The main reason attributed was soil degradation and the consequent deterioration of soil quality. The processes leading to soil degradation are generally triggered by excessive pressure on land to meet growing demands of rapidly increasing population for food, fodder and fiber. Various anthropogenic activities and over exploitation of natural resources without considering the ecological balance of the system have accelerated the soil degradation through the processes like salinization, acidification, waterlogging, drought, pollution etc. These processes in turn, reduce agricultural productivity. Though a large number of factors are associated with soil degradation, the major ones are presented below.

1. **Agricultural activities:** Agricultural activities and practices cause land degradation in a number of ways depending on land use, crops grown and management practices adopted. Some of the common causes of land degradation by agriculture include cultivation in fragile areas and marginal sloping lands without any conservation measures, land clearing through clear cutting and deforestation, agricultural depletion of soil nutrients through poor farming practices, excessive irrigation, excessive tillage, poor crop rotation, burning of crop residues, low organic input addition, over drafting (the process of extracting groundwater beyond the safe yield of the aquifer) and land pollution.
2. **Overgrazing, deforestation and careless forest management:** Overgrazing and deforestation have caused degradation in eight Indian states which now have >20% wasteland. Loss of vegetation occurs due to cutting beyond the silviculturally permissible limit, unsustainable fuelwood and fodder extraction, encroachment by agriculture into forest lands, forest fires and overgrazing, all of which subject the land to degradation forces.
3. **Industrialization, urbanization and infrastructure development:** An increase in industrialization, urbanization and infrastructure development is progressively taking away considerable areas of land from agriculture, forestry, grassland and pasture, and unused lands with wild vegetation.
4. **Mining:** Opencast mining is of particular focus because it disturbs the physical, chemical, and biological features of the soil and alters the socioeconomic features of a region. Negative effects of mining are water scarcity due to lowering of water table, soil contamination, part or total loss of flora and fauna, air and water pollution and acid mine drainage. Overburden removal from mine area results in significant loss of vegetation and rich topsoil.
5. **Natural and Social causes:** Natural causes of land degradation include earthquakes, tsunamis, droughts, avalanches, landslides,

volcanic eruptions, floods, tornadoes, and wildfires. Some underlying social causes of soil degradation are land shortage, decline in per capita land availability, economic pressure on land, land tenancy, poverty, and population increase.

Soil degradation is broadly defined as the physical, chemical and biological decline in soil quality, caused mainly due to its misuse by humans. It can be defined as the rate of adverse changes in soil qualities such as nutrient status, soil depth, soil reaction, salt content etc. resulting decline in productive capacity of land due to processes induced mainly by human intervention. It can be the loss of organic matter, decline in soil fertility and structural condition, erosion, adverse changes in salinity, acidity or alkalinity, and the effects of toxic chemicals, pollutants or excessive flooding. To assess the extent of soil degradation, degradation maps were prepared and in India degraded soils cover an area of 187 M ha, representing almost 57 per cent of the total geographical area (Bhattacharyya *et al.,* 2015).

Types of Degradation

1. **Physical degradation:** It involves physical loss and the reduction in quality of topsoil. The physical loss of soil is mainly due to soil erosion by water or wind. Tillage also encourages soil erosion. Some other processes that contribute to physical degradation are soil sealing, crusting, compaction, waterlogging and aridification.
2. **Chemical degradation:** It is defined as the accumulated negative impact of chemicals and chemical processes on those properties that regulate the life processes in the soil. A "healthy" soil has important physical, chemical and biological attributes including nutrient supply, acid and base buffer capacity, organic matter decomposition, pathogen destruction, toxic metal inactivation, and toxic organic inactivation and degradation. Mismanagement can disrupt the above function of a healthy soil and how these processes are affected decides the extent of chemical degradation. Processes such as acidification, salinization, sodification, soil pollution etc. are responsible for chemical degradation.

3. **Biological degradation:** It is the impairment or elimination of one or more "significant" population of microorganisms in soil, often with a resulting change in biogeochemical processes within the associated ecosystem. "Significant" microorganisms are those for which an ecologically significant role is understood. Soil microorganisms play key roles in cycling of nutrients, decomposition of wastes and residues, and detoxification of pollutant compounds in the environment. Factors that adversely affect these organisms, and/or their ability to mediate anthropic functions result biological degradation.

An unhealthy scenario exists between the developmental activities and the environmental preservation, as most of the former are not ecofriendly. Large areas of productive soils are being deprived of cultivation due to the resultant adverse effects. Even some of the production technologies like intensive cultivation, excessive irrigation, unscientific use of fertilisers and agrochemicals which are responsible for soil degradation have profound environmental implications.

It is difficult to part physical, chemical and biological degradation, since they are so intertwined. If a soil is physically degraded naturally its chemical and biological activities are at very low rate. The same is the case with chemical and biological degradation. Hence it is better to have a description of various types of degraded soils and their reclamation measures.

1. ERODED SOILS

Soil erosion is a complex phenomenon governed by several factors and causes tremendous damage to about one-third of world's cropped lands. Though natural geologic erosion can take place, the rate of erosion has assumed disastrous proportions due to intensification of various human induced activities. Soil erosion either water or wind, removes the fertile surface soil exposing the subsurface strata. If control measures are not taken, the subsoils are also subjected to erosion. Both, the areas from where the soil is removed and where it is deposited are adversely affected by erosion. The transported soils are usually get deposited in reservoirs and reduces their water holding capacity. According to a 2015 report of the Indian Institute of Remote Sensing, the estimated amount of soil erosion that occurred in India was 147 M ha. Under this broad figure, 94 M ha were claimed by water erosion, 16 M ha by acidification, 14 M ha by flooding and 9 M ha from wind erosion, 6 M ha from salinity and 7 M ha from a combination of factors. 29 percent of the soil that is eroded is lost in the sea while 61 percent is just relocated (Bhattacharyya *et al.*, 2015).

The intensity of soil erosion is rated based on the amount of soil removed per hectare per year. If it is < 5 t ha^{-1} yr^{-1}, it is rated as slight; 5 - 10- moderate; 10-20 high; 20-40 very high; 40-80 severe and >80 t ha^{-1} yr^{-1} as very severe. According to the National Bureau of Soil Survey and Land Use Planning (2004) about 146.8 M ha is degraded. Water erosion is the most serious degradation problem in India, resulting in loss of topsoil and terrain deformation. Based on first approximation analysis of existing soil loss data, the average soil erosion rate was 16.4 t ha^{-1} yr^{-1}, resulting in an annual total soil loss of 5.3 billion tons throughout the country (Singh, 2009). Nearly 29 % of total eroded soil is permanently lost to the sea, while 61 % is simply transferred from one place to another and the remaining 10 % is deposited in reservoirs.

Wind erosion is more prominent in areas having sandy soil with low organic matter content. Removal of vegetative covers and overgrazing enhances the intensity and extent of wind erosion. Wind erosion is the major cause behind desertification. The sand movement causes colossal damage to the adjoining cultivated areas, roads, canals, buildings etc. Sand dunes of varying heights are encountered, but get stabilized with the establishment of good cover of grasses, bushes and other vegetation. In India, wind erosion is at its worst form in Western Rajasthan. Its impact reaches up to Delhi and nearby Himalayan tracts.

Characteristics of Eroded Soils

Erosion removes a field's original topsoil, exposing the subsoil materials and is the principal cause of the loss of organic matter and nutrients in soils, which undoubtedly contributes to the loss of productivity. These soils are characterized by the following features.

1. **Shallow soils or compacted soils with A-C type profiles:** Removal of surface soil exposes the subsoil and decreases the soil depth. The raindrop impact and drying of soils often initiates soil compaction and reduce the soil productivity.
2. **Weak structure:** The breakdown of aggregates and the removal of smaller particles or entire layers of soil or organic matter weaken the structure. The soil structure is likely to be coarser, less stable and subject to more damage by rainfall impact, tillage or traffic.
3. **Changes in soil texture:** The removal of top soil and organic matter and intermixing with subsoil changes the soil texture. Textural changes can in turn affect the water-holding capacity of the soil, making it more susceptible to extreme conditions such as drought.
4. **Low water holding capacity:** Loss of organic matter, clay and increase in bulk density decreases water holding capacity
5. **Low organic carbon:** Eroded soils are low in organic carbon content and their humic substances
6. **Poor fertility:** Soils become poor in fertility due to loss of surface soil rich in nutrients

Effects of Water Erosion

The implications of soil erosion by water extend beyond the removal of valuable topsoil. Crop emergence, growth and yield are directly affected by the loss of natural nutrients, organic matter and applied fertilizers. Seeds and plants can be completely removed by erosion. Pesticides may also be carried off the site with the eroded soil.

The off-site impacts of soil erosion by water are not always as apparent as the on-site effects. Eroded soil, deposited down slope, inhibits or delays the emergence of seeds, buries small seedlings and necessitates replanting in the affected areas. Sediment that reaches streams or watercourses can accelerate bank erosion, obstructs the natural water flow, get deposited in reservoirs, damage habitat of aquatic organisms and degrade downstream water quality. Pesticides and fertilizers, frequently transported along with the eroding soil, contaminate or pollute downstream water sources, wetlands and lakes.

Effects of Wind Erosion

Wind erosion damages crops through sandblasting of young seedlings or transplants, burial of plants or seeds and exposure of seeds. Plants damaged by sandblasting are more vulnerable to diseases and results a decline in yield and quality of produce. Also, wind erosion can create adverse operating conditions, preventing timely field activities.

Soil drifting is a fertility depleting process that can lead to poor crop growth and yield reductions in areas of fields where wind erosion is a recurring problem. Continual drifting in an area gradually causes a textural change in the soil. Loss of fine sand, silt, clay and organic particles from sandy soils lower the moisture-holding capacity of the soil and increases soil erodibility. Also, soil nutrients and surface-applied chemicals can be carried along with the soil particles, contributing to off-site impacts. In addition, blowing dust can affect human health and create public safety hazards.

In nutshell the consequences of erosion are

1. Loss of top soil and fertility
2. Loss of soil nutrients and organic matter
3. Loss of vegetation and habitat
4. Lowering of ground water level

5. More prevalence of vagaries of climate change like frequent drought, flood, drying of rivers, landslides etc.
6. Adverse effect on national economy.

Conservation Measures / Management

Each soil will have an upper threshold limit of erosion termed as "Tolerance to soil loss (T)" that can be allowed without degrading long term productivity. If soil erosion rates are greater than T, mitigation measures are needed to achieve sustainable productivity. The adoption of various soil conservation measures reduces soil erosion by water and wind. Tillage and cropping practices, land management practices and water harvesting techniques directly affect the overall soil erosion problem and solutions on a farm. When crop rotations or changing tillage practices are not enough to control erosion on a field, a combination of approaches or more extreme measures might be necessary. For example, contour plowing, strip-cropping or terracing may be considered. In more serious cases where concentrated runoff occurs, it is necessary to include structural controls like grassed waterways, drop pipe and grade control structures, rock chutes, and water and sediment control basins etc.

Soil conservation measures, such as contour ploughing, bunding, use of strips and terraces can decrease erosion and slow runoff water. Mechanical measures like physical barriers such as embankments and wind breaks, or vegetation cover / geotextiles and soil husbandry are important measures to control soil erosion. In addition, conservation agriculture, agroforestry, integrated nutrient management and diversified cropping also conserve soil and water. Some of the most effective agronomic / engineering methods for soil conservation are briefed below.

1. **Strip cropping:** Strip cropping is a very effective and inexpensive method for controlling soil erosion. It is a combination of contouring and crop rotation in which alternate strips of erosion permitting (row crops) and resistant / tolerant crops are grown on the same slope, perpendicular to the wind or water flow. When soil is detached from

the row crops by the forces of wind or water, the dense soil conserving crops trap the soil particles and reduce wind translation and/or runoff.

2. **Mulching:** It is a type of protective covering which is in direct contact with the ground, provides protection to soil from water and wind erosion. It can vary from natural materials like straw, dried leaves, wood chips, saw dust etc. to synthetic mulches. Mulching materials are spread evenly onto the ground. Apart from reducing the rain fall impact, it helps to reduce water evaporation, regulate soil temperature and control weeds. Natural mulches enrich the soil through their decomposition.
3. **Contour bunding:** Bunds are constructed along the contour to reduce the velocity of flowing water. Bunds help to check the velocity of the run-off, to carry excessive rainfall safely downstream and to let off stream flow in natural channels. Bunding helps to retain the rainwater for more time and thereby allowing rainwater to percolate into the soil. It is better to adopt possible agronomic conservation measures on the constructed bunds so as to enhance the protection. Land slope and soil characteristics are considered for selection of bund type and design. Three types of bunds viz., earthern bunds, stone pitched contour bunds and graded bunds are usually constructed.

Earthern bunds are small embankment type structures made up of locally available earth materials. Stone pitched contour bunds are constructed in contour at suitable intervals in slopes. Stone pitched contour bunds are very suitable for laterite soils and is highly suitable for steep slopes, up to 35% .

Graded bunds are adopted in areas having low infiltration (< 8 mm h^{-1}) and more than 800 mm rainfall. Graded bunds are laid along pre-determined longitudinal grade instead of along the contours for safe disposal of excess runoff. Gradient given may vary from 0.4 to 0.8 % (0.4 % for light soils and 0.8 % for heavy soils).

4. **Vegetative hedges:** Runoff velocity can be reduced drastically by planting vegetative hedges, bunch grass, or shrubs on the contour at regular intervals.These hedges can increase the time for water to infiltrate into the soil and facilitate sedimentation and deposition of eroded material by reducing the carrying capacity of the overland flow. Vegetative hedges or narrow grass strips serve as porous filters. These hedges may not reduce runoff amount but can drastically decrease soil loss.

5. **Trenches:** Contour trenches are used both on hill slopes as well as on degraded and barren waste lands for soil and moisture conservation and afforestation purposes. The trenches break the slope and reduce the velocity of surface runoff. It can be used in all slopes irrespective of rainfall conditions, varying soil types and depths.

   a) **Contour trench:** Contour trenches are ditches dug along a hillside in such a way that they follow a contour and run perpendicular to the flow of water. Trenches are constructed as continuous, across the slope so that water flowing downhill is stopped in its tracks by the trenches, and water percolation into the soil below is facilitated. This will stabilize the soil around the trenches.

   b) **Staggered trench:** The length of trenches is kept short up to 2-3 m in a row along the contour with interspaces (5-7 m) between them. It is suited for medium rainfall areas with dissected topography.

6. **Terracing:** Terracing is a combination of contouring and land shaping in which earth embankments or ridges, are designed to intercept runoff water and channel it to a specific outlet. Terraces reduce erosion by decreasing the steepness and length of the hillside slope and by preventing damage done by surface runoff

   There are basically two types of terraces:

   1. Bench terraces
   2. Broad base terraces

The bench terrace, perhaps one of the oldest forms of terraces, is used to reduce land slope. The broad base terrace, on the other hand, is used to control and retain surface water on sloping land.

7. **Moisture conservation pits:** Any form of depression or micro pit is constructed over the land surface to arrest excess surface runoff and silting and thus leading to ground water recharge. Pits of suitable dimension are constructed in the field which would impound water and contribute to ground water recharge during rainy season. The silt accumulated in the pits could be dug out and used in the farmer's field which would improve nutrient status of the soil. Much care has to be given while taking pits in an area, since excessive number of pits results storage of moisture greater than the storage capacity of soils which may lead to the phenomenon of soil piping and finally give way to landslides.
8. **Agrostological measures:** Suitable grass species preferably fodder grass is planted in rows across the slope. It can also be planted on the berm of the bunds. The fibrous root system of grass will offer better protection of the top soil and filter the run off to trap the sediments. Well established grass species reduces erosion by moderating the impact of raindrops and also increases the infiltration opportunity time. The grass thus planted could be used as fodder for livestock, which could also be an alternate source of income to the farming community.
9. **Vegetated waterways:** Vegetated waterways are built to protect soil against the erosive forces of concentrated runoff from sloping lands. By collecting and concentrating overland flow, waterways absorb the destructive energy which causes channel erosion and gully formation.Waterways can have cross sections in parabolic, trapezoidal, or triangular form, depending on the functional requirements. Several of these requirements are climate, channel capacity and desired flow velocity.

   Grass linings should be hardy with dense - growing perennials adapted to the geographical region and soil. The grass should be cut periodically, fertilized as needed and not subjected to prolonged traffic by either livestock or vehicles.

Bio fencing along the banks of waterways or streams helps to protect the soil from water erosion and its further deposition in water bodies.

10. **Contouring:** Contouring refers to perform all tillage and planting operations of crops on or near the same elevation or "contour." It is applicable on relatively short slopes up to about 8 per cent steepness with fairly stable soils. By planting across the slope, rather than up and down a hill, the contour ridges slow or stop the downhill flow of water. Water is held in between these contours, thus reducing water erosion and increasing soil moisture. Contouring's impact on annual soil loss rates varies with slope steepness.

11. **Agroforestry**: Agroforestry measures consist of planting of woody perennials (trees, bamboo, shrubs etc.) wherever necessary to control soil erosion. These measures reduce erosive force of water through impeding effects of tree roots and through soil cover provided by the tree canopy and litter. These are potential enough to conserve soil and moisture in the area through a combination of mulching and shading

12. **Drainage line treatments:** Drainage channels / gullies are the carriers of runoff and sediment in watershed. Steep bed gradient (slope) of a channel cause high runoff velocity with associated heavy sediment flow. Hence channel gradient needs to be reduced in order to being the runoff velocities within permissible limits. The drainage line treatments aim at betterment of the drainage lines and facilitate better local production environment of the concerned areas by providing the required basic infrastructure. Moderation of floods and related damage, control of saline intrusion etc. are the major areas which facilitate increased productivity.

13. **Check dams:** Check dams are constructed across the streams to reduce the velocity of runoff water and to entrap the sediments. Check dams are built in a range of sizes using a variety of materials,

including clay, stone and cement. Earthen check dams, or embankments, can easily be constructed by the farmers themselves. There are logwood check dams, loose boulder check dams, dry rubble check dams and concrete / masonry check dams. Masonry and reinforced cement concrete (RCC) check dams are of more permanent in nature and serve the purpose of water conservation. The sluice, spillways and other regulatory structures constructed in the drains help to regulate the flow of excess water.

Loose boulder check dam is constructed using locally available stones and other materials across the stream to reduce runoff velocity and to entrap the sediments. Masonry check dams are constructed across streams of heavy flow and high velocity. The construction cost is usually high and requires hydrologic and hydraulic designs. Gabion check dams are very much suitable for degraded locations like high rainfall areas with torrential streams / drainage lines to stabilize.

14. **Stream bank stabilization:** Due to undulating topography and high intensity of rainfall, huge quantities of rainwater flow through the drains. The existing drains are silted and flow of water through the drains is restricted resulting in severe bank erosion, overflow and spreading of water through adjacent agricultural land causing damage to crops. Stream bank stabilization is done by constructing retaining walls of different configurations and design. By stabilizing the stream banks, the flow can be regulated, scouring of banks avoided and drainage congestion can be averted.

15. **Farm ponds and water harvesting structures:** Farm ponds are mainly meant for the purpose of storing the surface runoff. The farm ponds and water harvesting structures constructed in the low lying areas contributed to the conservation of excess rain water and the replenishment of ground water. The water harvesting structures are constructed as masonry structures. The impounded water also provides lifesaving irrigation to the lands in the ayacut. It will increase the soil moisture regime around the structure for increased crop production and recharge the ground water.

16. **Coir geo-textiles:** Coir geo-textiles is used as an erosion control measure for lake, canal and river bank protection. Coir mesh–mattings are used extensively in erosion control works. The ultimate objective is to establish a dense network of root system and vegetative cover to the desired degree of growth in the shortest possible time. Coir geo textiles intercept rainfall and aid *in situ* moisture conservation.
17. **Percolation ponds:** Percolation tanks are small water harvesting structures constructed across natural stream or water course to artificially recharge ground water during lean months. These structures increase the availability of water in the wells of surrounding area even during dry spells, which can be utilised by farmers for domestic as well as for irrigation purposes. They are constructed across the streams / canals for checking velocity of run off for increasing water percolation, improving soil moisture and promote siltation. Pineapple or guinea grass could be planted on the berm of the pond which will apart from providing better reinforcement; provide additional income by way of fodder / fruit. During the dry seasons, the ponds could be put to alternate use by dumping organic wastes which can be used as manure before the onset of monsoon.

Conservation Measures Against Wind Erosion

Many practices are used to reduce wind erosion, but all are basically directed at accomplishing one or more of the following:

1. **Reduce wind velocity at the soil surface:** This is done with windbreaks, crop residues, cover crops, surface roughness and strip cropping.
2. **Trap soil particles:** This is accomplished by maintaining crop residues on the soil surface and/or by ridging or roughening the soil surface.
3. **Increase size of soil aggregates:** Size of soil aggregates can be increased by management practices like crop rotations including grasses and legumes, growing high-residue crops and returning crop residues to the soil, proper manuring and emergency tillage (which can create stable clods on the soil surface if soil moisture and texture allow).

The successful wind erosion controlling measures are described below.

a. **Stubble mulching:** Stubble mulching is one of the effective ways to control wind erosion and conserve soil moisture. This practice is most appropriate with wheat and other small grains, and sorghum. The stubble mulch is particularly needed on sandy soils where a rough surface cannot be maintained. Tillage practices which result in a rough cloddy surface, preferably with trash cover or stubble mulch help to check wind erosion. Rough tillage traps the saltating particles, decreases wind velocity at the soil surface, prevents soil particles from breaking up into small particles which will drift, and increases percolation rate and decreases runoff.

   The quantity of crop residue required to control wind erosion can be estimated from the analysis of the major factors that affect soil blowing. For example, the more susceptible the soil is to movement by wind, the more residues are required to prevent it from blowing. Large fields require more residue than narrow fields or fields protected by wind breaks and shelter belts. Vegetables and other crops that are damaged by abrasion require more residue than do field crops. Arid areas need more residue than humid areas, the regions of high winds require more cover than those of low winds.

b. **No tillage or minimum tillage:** Crops are planted directly into the residue of the previous crop. Thus, more residues can be conserved for wind erosion control if tillage operations could be avoided. As effective chemicals for weed control are developed, no tillage systems are increasingly used. Herbicides are also combined with limited tillage to provide better weed control and, at the same time, conserve as much residue as possible for effective wind erosion control. The effectiveness of herbicides and the techniques for applying them vary with climatic and soil conditions.

c. **Cover Crops:** A cover crop is any crop planted solely to control erosion. It is usually planted for protection when regular crops are off the land, but also may be planted in strips or between rows to provide protection for vegetables or other crops highly susceptible to abrasive injury in the seedling stage. Cover crops are usually grown between regular crops like wheat, corn, soybean etc. Examples of cover crops are annual rye grass, crimson clover, oats, oilseed radish, and cereal rye. Cover crops are more suited to humid areas because in drier areas they compete for moisture. In drier areas, the cover crops are used only for control on erosion-susceptible knolls, on land without protective cover, or on cultivated land.

d. **Strip cropping:** Strip cropping aids in the control of soil blowing by shortening the distance that loose soil can move. Strips laid out on contour lines also conserve moisture which in turn helps to prevent blowing. To control wind erosion, crop strips are run straight and at right angles to the prevailing winds. Strip cropping does not require any change in cropping practices, and it does not remove any land from cultivation. The field is simply sub-divided into alternate strips of erosion resistant crops and erosion susceptible crops or fallow. Erosion resistant crops include small grains and other crops seeded closely to cover the ground rapidly. Erosion susceptible crops are tobacco, sugar beets, peas, beans, potatoes, peanuts, asparagus, etc. Cotton and sorghum are intermediate in resistance to wind erosion. Crops like vetch, rye and clover are erosion resistant crops.

   Strip cropping controls soil blowing by reducing soil avalanching. The rate of soil avalanching varies directly with the erodibility of the soil and the width of the eroding field. So one factor that determines the width of the strips is the kind of soil and another is land use. In the vegetable growing areas, for example, buffer strips consisting of very narrow strips of rye, wheat, or grass may be used with wider strips on erosion susceptible land. A common practice is to make the buffers one-tenth as wide as the erosion susceptible strips. In drier areas with wheat and sorghum, erosion resistant and erosion susceptible strips are generally equal in width.

2. NUTRITIONALLY POOR SOILS

Most of the soils have nutritional problems either being deficient in one or more of the essential nutrients or the excess of them. A particular soil cannot be rated as with nutritional problems without the support of soil analysis data. However, we know certain soils viz., sandy soils, laterite soils, red soils etc. are poor in soil fertility due low content of most of the essential nutrients and organic matter. At the same time acid sulphate soils, sodic soils, saline soils etc. may have the problems of excess of some of the elements. Hence a general categorization of soils based on their deficiency or excess / toxic quantity is not possible. Since the general nutrient deficiencies and toxicities are beyond the purview of this book, they are not included here.

3. WATERLOGGED SOILS

Extensive waterlogging caused by the rise in water table pose a great threat to soil productivity and environmental ecology. Nearness to sea, receipt of heavy rainfall and poor drainage and faulty irrigation are some of the major reasons for the development of waterlogged soils of which the last two are mainly anthropogenic. In irrigated areas, continued increase in irrigation potential had damaged the soil quality and even sets in secondary salinization.

In India, complaints regarding deterioration of soil quality due to irrigation were reported as early as in 1855 after the commissioning of Western Yamuna canal in 1839. According to NBSS and LUP, around 14.3 M ha of land was affected by waterlogging (NBSS and LUP, 2004). There is ample evidence that the fertile lands are affected by waterlogging within few years after introduction of irrigation. Waterlogging has taken place in almost all irrigation commands in arid and semi-arid tracts of

India. Secondary salinization was noted in north-west regions of Rajasthan and south-west parts of Haryana, due to poor quality irrigation water. Waterlogging is also observed in depressions, backwater swamps, tidal flats, deltas, mudflats, and estuaries. These lands are also put to heavy siltation and pollution from industries and other anthropogenic activities.

4. CONTAMINATED SOILS

The anthropogenic activities had increased the pollution of soils all over the world. The soils get contaminated by different types of materials released from the agricultural, industrial and development activities. The major contaminants include agrochemicals, industrial effluents and automobile emissions. From the agriculture sector the major contaminants are fertilizer and pesticide residues.

a. *Inorganic fertilizers*

The success of sustainable food production greatly relies on the widespread use of mineral fertilizers. Large quantities of fertilizers are regularly added to soils in intensive farming systems to provide adequate N, P, K and other essential nutrients for crop growth. Only a fraction of the added nutrients is taken up by the plant and the rest gets either accumulated in soil or subjected to runoff loss. Precise application of fertilizers to the plants gives promising results, however, the same fertilizers when used indiscriminately, pose serious threat to the environment by polluting soil and water bodies. Therefore, it is highly desirable to understand the factors responsible for the fertilizer-induced contamination of soils and water system; and also, to innovate ways that could effectively check the spread of this nonpoint source pollution without compromising growth and yield of crop plants.

Nonpoint source pollution generally results from many diffuse sources like land runoff, precipitation, atmospheric deposition, drainage, seepage or hydrologic modification. It includes excess fertilizers, herbicides and insecticides from agricultural lands and residential areas, oil, grease and toxic chemicals from urban runoff and energy production, sediment from improperly managed construction sites, crop and forest lands, and eroding streambanks. Nonpoint source pollution is caused by rainfall or snowmelt moving over and through the ground. As the runoff moves, it picks up and carries away natural and human-made pollutants, finally depositing them into lakes, rivers, wetlands, coastal waters and ground waters.

The nutrients present in fertilizers get easily mobilized by rainfall and reach the surface water which result in the nutrient enrichment of surface water bodies leading to a condition known as eutrophication. Presence of excess nutrients especially N and P allows luxurious growth of aquatic plants and algal blooms that causes depletion of dissolved oxygen which negatively affects aquatic life and ecosystem. The soluble forms of these nutrients supplied from fertilizers dissolve in water and are transported in solution through deep percolation of irrigation and rainfall water to the groundwater where they make potable water supplies unsuitable for humans and livestock consumption. Nitrogen-based fertilizers are considered the worst culprits of water pollution since nitrate being the final product of N mineralization which gets easily leached. Nitrate levels above 10 mg L^{-1} as nitrogen or 45 mg L^{-1} as nitrate ion in the groundwater can cause methemoglobinemia (blue baby syndrome). The nutrients, especially nitrates in fertilizers can cause problems for natural habitats and for human health if they are washed off soil into watercourses or leached through soil into groundwater.

Another threat from nitrogenous fertilisers is soil acidification. Nitrogen from applied fertilizers gets oxidized to nitrates and produces protons and decreases soil pH that results in the acidification of soil. Acidified soil causes mineral toxicity and mineral deficiency that ultimately results in soils with poor health.

The compounds used to supply essential nutrients especially P contain trace amounts of heavy metals Cd, Pb, As, Cr, Zn, Th etc. and small quantities of radionuclides as impurities, which, after continued fertilizer application may significantly increase their content in the soil. The contaminants in phosphatic fertilizers owe their existence to its origin as almost all of the world's phosphate fertilizers are derived from phosphate rocks. These metals (Cd, As and Pb) are of most significant concern as they adversely affect human health and cause musculoskeletal problems.

The deleterious effect of the chemical fertilizers will itself start from the manufacturing of these chemicals, whose products and byproducts are some toxic chemicals or gases like NH_4, CO_2, CH_4 etc. which will cause air pollution. And when the wastes from the industries are disposed untreated into nearby water bodies, it will cause water pollution.

The adverse effect of these synthetic chemicals on human health and environment can only be reduced or eliminated by adopting new agricultural technological practices such as integrated nutrient management or good agricultural practices which include the use of organic inputs such as manure, biofertilizers, biopesticides, slow release fertilizers and nanofertilizers etc. and their use in optimum quantity and proper time. This would improve the application efficiency as well as use efficiency of the fertilizers. Application of fertilisers should strictly be based on soil test data which will help to maintain a healthy nutrient balance in soil. Modern methods of irrigation with water saving and techniques like precision farming could reduce nutrient loss.

Therefore, to cope with the adverse effects of fertilizers and the contaminants present, it is necessary to develop crop varieties that utilize the available nutrients efficiently. The future research should be directed to develop fertilizers with minimum contaminants and to explore the optimum dose of fertilizers for a particular crop with minimum losses to the environment. Fertilizer-induced contamination of soil and water can also be minimized by adopting various control measures such as phytoremediation, application of lowest possible dose of fertilizers, wastewater treatment, nutrient monitoring, and fertilizer application based on mathematical models, creating public awareness, and imposing necessary legislations.

b. Organic fertilizers

Organic fertilizers are naturally occurring compounds produced from waste matter or byproducts, where only the physical extraction or processing steps are assisted by man. Commonly used organic fertilizers include composted animal manure, compost, sewage sludge, food processing wastes, and municipal biosolids. They improve soil health and release nutrients to soils gradually. However, the use of organic fertilizers also has many disadvantages such as excess availability of N compared to P and K, presence of heavy metals and other toxic substances. Use of animal manure as soil amendment and nutrient source has been a long tradition worldwide. However, recently the input of manure has been recognized as a major source of metal enrichment in soils.

c. Pesticides

Pesticides and herbicides applied to agricultural land to control pests and weeds respectively, disrupt crop production. Soil contamination can occur when pesticides persist and accumulate in soils, which can alter microbial processes, increase plant uptake of the chemical, and also cause toxicity to soil organisms. The extent to which the pesticides and herbicides persist, depends on the compound's unique chemistry, which affects sorption dynamics and resulting fate, and transport in the soil environment.

Pesticide leaching occurs when pesticides mix with water and move through the soil, ultimately contaminates the ground water. The amount of leaching is correlated with soil texture and pesticide characteristics and the degree of rainfall and irrigation. Leaching is most likely to happen if using a water-soluble pesticide, when the soil tends to be sandy in texture; if excessive watering occurs just after pesticide application; or if the adsorption ability of the pesticide to the soil is low. Leaching may not only originate from treated fields, but also from pesticide mixing areas, pesticide application machinery washing sites, or disposal areas.

Naturally the pesticides reached the soil or water get degraded. But it depends on the chemical nature of the pesticides and its degradation pattern. The degradation can be by light, heat and by microorganisms. The microbial ability of degrading the pesticides is widely employed in bioremediation techniques especially for organic pesticides. Some of the pesticides contain heavy metals, which may reach soil or water when applied. The heavy metal problems can be addressed by bioremediation and/or phytoremediation that uses microbes, vegetation and associated microbiota, soil amendments, and agronomic techniques to remove, contain, or render environmental contaminants harmless.

Scientific and judicious use of pesticides especially the new generation pesticides along with efficient application techniques could address these problems to a certain extent.

5. MINED DEGRADED SOILS (MINE SOILS)

Surface mining for coal and other valuable geological materials have been causing severe ecological disturbances worldwide since pre-historic times. The rate of consumption of mineral resources is still increasing due to the acceleration of urbanization, population growth and advancement in technology and science, which has been exceptionally fast in the 20th and 21st century. Mining activities for mineral resources over the years have resulted in major soil damage. Due the removal process of desired mineral materials, soil textures have been destroyed, various nutrient cycles have been disturbed, and microbial communities have been altered, affecting vegetation and leading to the destruction of wide areas of land in many countries.

Mined degraded soils are man-made habitat which experience a wide range of problems for establishing and maintaining vegetation, depending on the types of mines such as metal mines, coal mines and quarries. The mining of ores and minerals results in deleterious environmental outcomes such as clearance of landscape biota and the production of large amount of by-products. During mining operations, initial vegetation clearance is followed by removal, and storage of the topsoil to expose the deeper mineral containing substrates. This topsoil is usually stored in a non-planted state for extended periods of time, often years. Over time, there are significant declines in key traits such as carbon content, seed banks of locally adapted native plants, and the composition of the microbiome including community diversity, and function.

Characteristics of Mined Soil

1. Most of the top soil will be lost and only skeletal materials remains
2. Excessively eroded and highly compacted soils
3. The landscape will be undulating / slopy with coarse textured dump materials ends up on the lower slopes
4. Soils are with low nutrient reserves but in some cases chemical extremes (Eg. acid rock drainage) exists
5. Very low water holding capacity
6. They are acidic or saline depending on the nature of materials left and the byproducts formed
7. Very low in organic matter content and major nutrients such as nitrogen and phosphorus
8. Excess of trace elements especially iron, manganese, copper etc. depending on the metals mined
9. Vegetative cover in these soils is very poor, with the exception of some grasses.
10. Soil temperature extremes may exist

Apart from the abiotic soil factors, the ecosystem faces certain biotic filters like excess herbivore (from high deer and elk populations), competition (from seeded agronomic grasses and legumes), propagule availability (with large disturbances); phytotoxic exudates (from weedy species) and facilitation and species interactions (adverse ecological interactions). Many of these may act together to limit recovery, so for instance, steep slopes and adverse textures may be found on waste rock dump slopes.

Rehabilitation of Mined Soils

To achieve ecological restoration of the mined degraded soils, several remediation techniques are available. The conventional technologies for rehabilitating metal contaminated sites are landfilling, soil washing, solidification/stabilization, electro kinetic treatment and soil excavation. But most of these techniques are impractical, especially for large tailings areas due to high cost, high energy consumption, complex protocols and possible secondary pollution. It is better to adopt mine land reclamation strategies that will take into account the betterment of soil structure, fertility, microflora, and nutrient cycling in order to recover the land as much as possible and to act as a self-sustaining ecosystem.

Establishment of a vegetative cover on the degraded soil is one of the best techniques. A vegetative cover with the choice of appropriate vegetation like grasses / pastures / trees over the soil is highly useful to stabilize the bare area and to minimize the pollution problem. Hence based on the physical, chemical and biological properties of degraded soils, suitable plants can be selected. Phytoremediators are highly useful for treating metal-contaminated soils, for stabilizing toxic mine spoils, and the removal of toxic metals from the spoils respectively. Soil amendments should be added to aid stabilizing mine spoils, and to enhance metal uptake accordingly. The phytoremediation techniques are cost-effective and environmentally sustainable.

After selecting the plants, thrust must be given for correcting the soil as well as topographical problems, so that they can survive under degraded situation.

Most serious problems are the coarse textured nature of soil and steep slopes. It will be difficult to establish vegetation on slopes but can be achieved with proper care. The materials at the top of the slope are fine textured and suitable for plant growth, but this area is continually bombarded by the materials from above. In the middle of the slope, the materials are too coarse to hold moisture and therefore will not support vegetation. At the bottom of the slopes, the spaces between the large rocks slowly fill with bryophytes and organic matter and eventually will support the growth of higher plants. The slow erosion of the fine textured materials from the top over the middle portion of the slope will finally allow vegetation to establish.

Another problem is the compactness of the soil. The use of excavators to prepare the surface of the compacted area can be an effective way of alleviating compaction. The excavator can make rough and loose surface configurations and can be used on re-contoured dump slopes to control erosion and provide suitable sites for vegetation establishment.

The low nutrient status of most mining wastes limits the growth of plants. However, the pioneering species, including those that are associated with nitrogen fixing bacteria can grow in areas of low nutrients and can be used as revegetation species on mining sites. Use of soil amendments and nutrient suppliers are to be encouraged if need arises.

If waste materials with adverse chemical properties (tailings and waste rock) are present, acid rock drainage and metal leaching will prevent the plant establishment. In such places waste rocks have to be covered with clean materials that are used to seal the wastes and allow vegetation to grow freely.

While rehabilitating the mined landscapes, along with the establishment of aboveground flora and fauna of adequate composition and diversity, the belowground microbiome also needs to be rehabilitated to produce stable ecosystem services. Plant growth promoting rhizobacteria (PGPR) are widely used to reclaim degraded soils. When these are used for mine site restoration, carefully consider preservation of native microbiome diversity through appropriate topsoil handling and storage.

Among the rehabilitation strategies, natural rebuilding processes provide the best available model for the restoration of degraded mined landscapes. To address the filters like compaction, steep slopes, adverse textures etc. sowing or planting of pioneering species have to be done to start the recovery processes. After the establishment of the pioneering species natural succession processes will ensure an appropriate vegetative cover on the degraded sites, building soils and replacing species as the ecosystems mature. By following the natural successional patterns in the establishment of vegetation, the right species will establish at the right time in the right place. By creating a diversity of habitats through the rough and loose surfaces, a diversity of species will establish. Creation of heterogeneous reclaimed areas will provide diverse ecosystems that will help to build resilience in the former mine site area.

BIBLIOGRAPHY

Arora, S., and Sharma, V. 2017. Reclamation and management of salt-affected soils for safeguarding agricultural productivity. *J. Safe Agri.* 1:1–10

Arora, S., Singh, Y.P., Vanza, M. and Sahni, D. 2016. Bioremediation of saline and sodic soils through halophilic bacteria to enhance agricultural production. *J. Soil Water Conserv.* 15:302–305

Arora, S., Vanza, M., Mehta, R., Bhuva, C. and Patel, P. 2014. Halophilic microbes for bioremediation of salt affected soils. *African J. Microbio. Res.* 8:3070–3078

Bhattacharyya, R., Ghosh N.B., Mishra P.K., Mandal B., Rao C.S., Sarkar D., Das, K., Anil, K.S., Lalitha, M., Hati, K.M. and Franzluebbers A.J. 2015. Soil degradation in India: challenges and potential solutions. *Sustainability* 7:3528-3570

Blakemore, L.C, Searle, P.L. and Daly, B.K. 1987. *Methods for Chemical Analysis of Soils*. New Zealand Soil Bureau, Lower Hutt, N.Z.

Buringh P. 1970. *Introduction to the Study of Soils in Tropical and Subtropical Regions*. Centre for Agricultural Publishing and Documentation, Wageningen (Netherlands), p 99

Chenery, H.B. 1954. Interregional and international input output analysis, the structure interdependence of economy. In: (ed) Barna, T. *Proceeding of an International Input Output Analysis Conference*, Gruffer, New York, Milano.

Dagar, J.C. and Minhas, P. 2016. *Agroforestry for the Management of Waterlogged Saline Soils and Poor Quality Waters*. Springer India, p 201

Fageria, N.K., Baligar, V.C. and Clark, R.B. 2002. Micronutrients in crop production. *Adv. Agron.* 77:185-268.

FAO, AGL 2000. Extent and causes of salt affected soils in participating countries. (http://www.fao.org/ag/agl/agll/spush/topic2.htm)

Gajbhiye, K.S. and Mandal, C.2006. Agro-ecological zones, their soil resource and cropping systems. In: (ed) IASRI. *Status of Farm Mechanization in India*, ICAR, New Delhi, pp 1–32

Goldhaber, M.B. and Kaplan, I.R.1974. Mechanisms of sulfur incorporation and isotope fractionation during early diagnosis in sediments of the Gulf of California. *Marine Chem.* 9:95-143

Kamprath, E. J. 1984. Crop response to lime on soils in the tropics. *In: (ed)* Adams, F. *Soil Acidity and Liming*, Am. Soc. of Agron., Madison, pp 349–368.

Kellogg, C.E. 1949. Preliminary suggestions for the classification and nomenclature of great soil groups in tropical and equatorial regions. *Comm. Bur. Soil Sci. Tech. Comm.* 46:76-35.

Latha, M.R. and Janaki, P. 2015. Problem Soils and their Management. http://www.agritech.tnau.ac.in/agriculture/agri_reosurcemgt_soil_soil contraints.html

Mandal, A.K., Sharma, R.C, Singh, G. and Dagar, J. 2010. Computerized database on salt affected soils in India. Tech. Bull. No.2. CSSRI, Karnal, pp 28

Mandal, K.G., Hati, K.M., Misra, A.K., Bandyopadhyay, K.K. and Tripathy, A.K. 2013. Land surface modification and crop diversification for enhancing productivity of a Vertisol. *Int. J. Plant Prod.* 7:455–472

Mandal, S., Raju, R., Kumar, A., Kumar, P. and Sharma, P.C. 2018. Current status of research, technology response and policy need of salt-affected soils in India – a review. *Ind. Soc. Coastal Agri. Res.* 36: 40–53

Martin F.J. and Doyne, H.C. 1927. Laterite and lateritic soils in Sierra Leone. *J. Agric. Sci* 17:530-547

Moore, P.A. and Patrick, W.H. Jr. 1989. Calcium and magnesium availability and uptake by rice in acid sulphate soils. *Soil Sci.Soc. Am. J.* 53:816-822

NBSS and LUP, 2004. National Bureau of Soil Survey and Land Use Planning. Soil Map (1:1 Million Scale); NBSS&LUP: Nagpur, India.

Pasricha, N.S., Nayyar, V.K., Randhawa, N. and Sinha, N.K. 1977. Influence of sulphur fertilization on suppression of molybdenum uptake by berseem (*Trifolium alexandrinum* L.) and oats (*Avena sativa* L.) grown on a molybdenum toxic soil. *Pl. Soil* 46:245-250

Ponnamperuma, F.N. 1972. The chemistry of submerged Soils. *Adv. Agron.* 24:29-96

Ponnamperuma, F.N. 1977. Behaviour of minor elements in paddy soils. *IRRI Research Paper Series* No. 8, IRRI, Los Banos, Laguna, Philippines, p 15

Pons, L.J., van Breemen, N. and Driessen, P.M. 1982. Physiography of coastal sediments and development of potential soil acidity. *Soil Sci. Soc. Am., Special Publ.* No. 10, pp 1–18

Pravalie, R. 2016. Drylands extent and environmental issues: A global approach. *Earth Sci. Rev.* 161: 259–278

Qadir, M., Quillerou, E., Nangia, V., Murtaza, G., Singh, M. and Thomas, R.J. 2014. Economics of salt-induced land degradation and restoration. *Nat. Res. Forum.* 38:282–295.

Qadir, M., Oster, J.D., Schubert, S., Noble, A.D. and Sahrawat, K.L. 2007. Phytoremediation of sodic and saline-sodic soils. *Adv. Agron.* 96: 197–247

Robinson G. W.1949. *Soils, their Origin, Constitution and Classification.* 3rd edn. London. Th. Marby, p 573

Scofield, C.S. 1940. Salt balance in irrigated areas. *J. Agric. Res*. 61:17-39

Sehgal, J. 2005. *A Textbook of Pedology: Concepts and Applications*. 2nd edn. Kalyani Publishers. New Delhi, India, p 514

Sehgal, J. and Abrol, I.P. 1994. *Soil Degradation in India: Status and Impact.* Oxford and IBH, New Delhi, India, p 80

Shahid, S.A. 2013. Developments in salinity assessment, modeling, mapping, and monitoring from regional to submicroscopic scales. In: (eds) Shahid, S.A., Abdelfattah, M.A., Taha, F.K. *Developments in Soil Salinity Assessment and Reclamation – Innovative Thinking and Use of Marginal Soil and Water Resources in Irrigated Agriculture*. Springer, New York, pp 3–43

Shahid S.A., and Rahman, K.R. 2011. Soil salinity development, classification, assessment and management in irrigated agriculture. In: (ed) Passarakli, M. *Handbook of Plant and Crop Stress*. CRC Press/Taylor & Francis Group, Boca Raton, pp 23

Sharma, D., Singh, A. and Sharma, P. 2015. Vision-2050. CSSRI, Karnal

Shoemaker, H.E., Mclean, E.O. and Pratt P.F. 1961. Determining lime requirement of soils with appreciable amount of aluminium. *Soil Sci. Soc America J.* 254: 274-277

Singh, G. 2009. Salinity related desertification and management strategies: Indian experience. *Land Degrad. Dev.* 20:367–385

Soil Survey Staff, 1975. *Soil Taxonomy, a Basic System of Classification for Making and Interpreting Soil Surveys*. USDA -NRCS, Washington DC, p 436

Varghese, T. and Byju, G. 1993. *Laterite Soils.* STEC, Government of Kerala, Thiruvananthapuram. p 116 van Breemen N. 1982. Genesis, morphology and classification of acid sulfate soils in coastal plains. In: (ed) Kittrick, J.A., Fanning, D.S. and Hossner, L.R. *Acid Sulphate Weathering* Vol.10. Wiley Online Library